Abdullah A. Shaikh
Heterogeneous Catalysis

Also of Interest

Aqueous Mediated Heterogeneous Catalysis
Edited by Asit K. Chakraborti and Bubun Banerjee, 2022
ISBN 9783110738452, e-ISBN 9783110733846

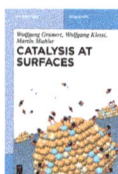

Catalysis at Surfaces
Wolfgang Grünert, Wolfgang Kleist and Martin Muhler, 2023
ISBN 9783110632477, e-ISBN 9783110632484

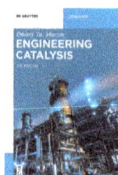

Engineering Catalysis
Dmitry Yu. Murzin, 2020
ISBN 9783110614428, e-ISBN 9783110614435

Process Engineering.
Addressing the Gap between Study and Chemical Industry
Michael Kleiber, 2020
ISBN 9783110657647, e-ISBN 9783110657685

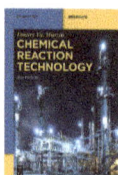

Chemical Reaction Technology
Dmitry Yu. Murzin, 2022
ISBN 9783110712520, e-ISBN 9783110712551

Abdullah A. Shaikh

Heterogeneous Catalysis

Solid Catalysts, Kinetics, Transport Effects, Catalytic Reactors

2nd Edition

DE GRUYTER

Author
Abdullah A. Shaikh
Chemical Engineering Department
King Fahd University of Petroleum and Minerals
Dhahran 31261
Kingdom of Saudi Arabia
aashaikh@kfupm.edu.sa

ISBN 978-3-11-103248-1
e-ISBN (PDF) 978-3-11-103251-1
e-ISBN (EPUB) 978-3-11-103300-6

Library of Congress Control Number: 2023938292

Bibliographic information published by the Deutsche Nationalbibliothek
The Deutsche Nationalbibliothek lists this publication in the Deutsche Nationalbibliografie;
detailed bibliographic data are available on the Internet at http://dnb.dnb.de.

© 2023 Walter de Gruyter GmbH, Berlin/Boston
Cover image: Bim/E+/Getty Images
Typesetting: Integra Software Services Pvt. Ltd.
Printing and binding: CPI books GmbH, Leck

www.degruyter.com

Preface to the second edition

Everywhere I go, I find a poet has been there before me.

In mid-2022, the publisher asked me if I was interested in preparing a new edition of the textbook. I enthusiastically welcomed the opportunity to revise and expand the original material, especially since I completed the first edition during the pandemic lockdown in the early months of 2020, under attendant limitations.

I have therefore revised almost all the chapters: some of the original artworks have been improved or replaced by better figures; materials here and there that were rather peripheral have been deleted; and some of the sections have been streamlined and expanded to provide clarity, depth, or both.

I have also added new examples and included solutions in the text. The problems in the last part of the text have likewise been revised and expanded.

However, new aspects and sections, which are not developed explicitly in the text, will be available on the textbook website for those readers who wish to expand their knowledge or look at details of the mathematical background of some design parameters and equations. The website will include, for example, sections on *catalyst preparation methods, transport-line reactors*, and *nonconventional industrial applications.*

It is fitting to reemphasize that my main ambition in this relatively brief textbook is to introduce the broad field of heterogeneous catalysis and its industrial applications. Therefore, this edition in spirit follows the original, in that focus is placed on the "essentials" that the graduating process engineer (and industrial chemist) would need as fundamental knowledge in the field. As I mentioned in the first edition, the text has enough material to provide a solid foundation for students who aspire to complete graduate studies in chemical engineering or industrial chemistry.

I welcome and look forward to feedback from the readers and instructors.

Abdullah A. Shaikh
Professor Emeritus
April 2023

https://doi.org/10.1515/9783111032511-202

Preface to the first edition

La simplicité fait la beauté

There are many books on heterogeneous catalysis, so it is appropriate to explain the motivation for writing the present book. All the existing books are excellent, and I have used many as references here and elsewhere. Yet I feel that most do not suit the background and language capabilities of undergraduate students in chemical engineering (and industrial chemistry), especially nonnative English speakers. All those books were written by well-known experts; however, in a broad sense, most (a) are too advanced, (b) address specific topics in heterogeneous catalysis, (c) are comprehensive or encyclopedic in nature, or (d) written in a language style that is overly complex for undergraduate students.

The present textbook is therefore not meant to be comprehensive, nor is it meant to give detailed coverage of solid catalysis and catalytic reactors. Rather, it is intended as an *introduction* to the vast field of heterogeneous catalysis and its industrial applications. The nine chapters contain sufficient material to be covered in a single semester. I hope they will provide the graduating engineer and industrial chemist, whether he/she intends to work in the chemical process industries or pursue graduate studies, with a "foundation" upon which he/she can build and expand his/her horizons.

These chapters grew from a set of lecture notes that I used in a senior undergraduate elective course on heterogeneous catalysis and its applications. Normally I cover most of the material in the nine chapters in about 13 weeks and leave the last 2 weeks for oral presentations of the term projects. I assign individual term projects that typically cover domestic industrial applications.

Most (chemical engineering) seniors taking this course have previously taken the first reaction engineering course, which means that some knowledge of chemical kinetics, thermodynamics, transport, and numerical methods is assumed.

Modeling and simulation are essential parts of what design engineers and R&D experts have been using in the last few decades. In the age of *IoT* and *Industry 4.0*, process engineers and industrial chemists are required to understand the powerful outcomes that modeling and simulation can capture.

Needless to say, numerical codes are essential for tackling realistic problems in catalytic kinetics and reactor analysis. Today, engineering students are trained in the application of a mainstream software package (e.g., MATLAB© and Mathcad©) to handle, at least, nonlinear algebraic equations and initial-value problems. I have included computational simulators and a few case studies in the last chapter. I hope to upload more numerical codes and process simulators for specific applications on the textbook's website.

https://doi.org/10.1515/9783111032511-203

As students and instructors know, problem-solving is an essential element in learning. The reader will find a relatively good number of problems at the end of the textbook covering a variety of topics in solid catalysis.

I welcome and look forward to feedback from the readers.

Abdullah A. Shaikh
Professor Emeritus
November 2019

Acknowledgments

I am grateful to the late Professor James J. Carberry of the University of Notre Dame (IN, USA). He was a great gentleman, mentor, and friend. He first introduced me to the fascinating field of catalytic reactor engineering during my graduate studies at Notre Dame. Those years were the most intellectually exciting time of my academic career.

My gratitude to the King Fahd University of Petroleum and Minerals for granting me a sabbatical leave during which a substantial portion of the book was written. Staff members at the KFUPM Research Institute, especially Dr. Abdullah M. Aitani and Dr. Ziauddin Qureshi, are gratefully acknowledged for providing industrial information, relevant photographs, drawings, and references. Dr. Abdulrahman S. Al-Ubaid and Dr. Nabeel S. Abo-Ghander are also acknowledged for their tangible and intangible support over the years.

The feedback and critique from some of my colleagues who have used my lecture notes to teach over the past decade, as well as my students' comments, are appreciated.

https://doi.org/10.1515/9783111032511-204

Contents

1 Introduction

Let the students ask questions!
A.P.J. Abdul Kalam

Most of us probably begin to learn about catalysts in high school. The following definition is typical of what we find in high-school chemistry textbooks or the internet: a catalyst is *"a substance that increases the rate of chemical reaction but is itself not consumed [in the process]."*

As college freshmen/women, we go on to learn other important observations related to catalysts, such as *"a catalyst changes only the rate of reaction but does not affect the chemical equilibrium"*; and *"a catalyst affects the rate of reaction by promoting a different molecular path for the reaction."*

Catalysis is defined as the occurrence, development, and use of catalysts and catalytic processes.

Catalysis can be divided into three categories: homogeneous, heterogeneous, and enzymatic catalysis. In *homogeneous* catalysis, the catalyst and reactants are in the same phase, usually a gas or a liquid; while in *heterogeneous* catalysis, a phase boundary separates the catalyst (usually a solid) from the reactants. In *enzymatic* catalysis, enzymes are substances (usually proteins) that play the role of catalysts in human beings and other living systems. Let us now look at some examples of these three categories.

An example of homogenous catalysis is the gas-phase oxidation of sulfur dioxide in the presence of nitrogen oxide:

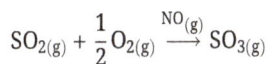

$$SO_{2(g)} + \frac{1}{2}O_{2(g)} \xrightarrow{NO_{(g)}} SO_{3(g)}$$

Another example is the decomposition of hydrogen peroxide over an aqueous bromide catalyst:

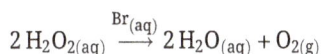

$$2\,H_2O_{2(aq)} \xrightarrow{Br_{(aq)}} 2\,H_2O_{(aq)} + O_{2(g)}$$

The Haber–Bosch process for the manufacture of ammonia, one of the most highly produced *commodity* chemicals, involves a heterogeneously catalyzed reaction, which proceeds as follows:

$$3H_{2(g)} + N_{2(g)} \rightleftharpoons 2NH_{3(g)}$$

The solid catalyst is iron-based such as magnetite (Fe_3O_4). The catalyst is the brown stone-like particles shown in Figure 1.1.

The production of methyl-tertiary-butyl ether (MTBE), an important fuel additive that is mixed with gasoline to increase the *octane number* and thus reduce automobile air pollution, is another example of a heterogeneously catalyzed liquid-phase reaction between *iso*-butylene and methanol:

https://doi.org/10.1515/9783111032511-001

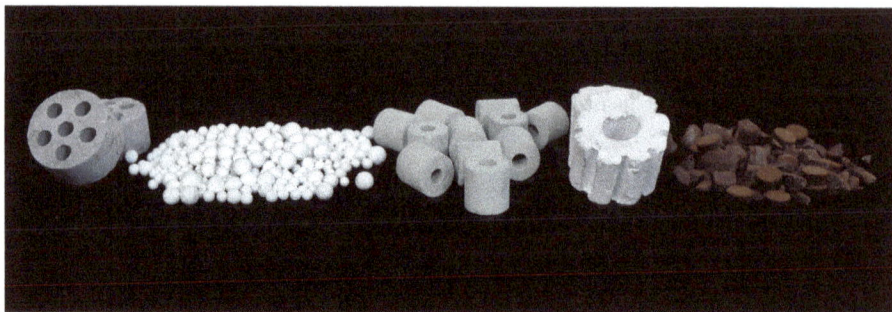

Figure 1.1: Different types and shapes of industrial catalysts.

$$CH_2 = C - (CH_3)_2 + CH_3OH \rightarrow CH_3 - O - C - (CH_3)_3$$

In this case the catalyst is an acidic ion exchange *resin*.[1] Figure 1.2 shows a picture of an industrial resin catalyst; the spherical resin particles are approximately 1 mm in diameter.

Figure 1.2: Typical spherical particles of a resin catalyst.

The synthesis of aspartame, a low-calorie sweetener (that some of us probably know under trade names such as NutraSweet™, Canderel™, and Hermesetas Gold™) is an example of an enzyme-catalyzed reaction. Aspartame is made from the combination of two amino acids, namely phenylalanine and aspartic acid in the presence of an enzyme.

1.1 Industrial importance of heterogeneous catalysis

This book deals specifically with heterogeneous catalysis. The majority (around 85% in some estimates) of industrial processes in petroleum-refining, petrochemical, and chemical industries are based on solid-catalyzed reactions.

---

1 A resin is (mostly) a synthetic semi-solid amorphous substance; resins are usually transparent yellowish to brown in color.

The important advantages for using solid catalysts include the following:

(a) The relatively easy separation between reactants, products, and solid catalyst, allowing the recycling of the catalyst to – or easy replacement in – the reactor.

(b) The relatively easy handling of solid catalysts.

(c) The possibility of "tailoring" the shape and properties of solid catalysts for specific purposes.

To further illustrate the importance of heterogeneous catalysts in industry, let us now consider Figure 1.3, which shows the *major* industrial applications of solid catalysts.

This partially explains why high budgets for R&D[2] are spent by giant chemical and petroleum companies, such as BASF, Dow-Dupont, Sinopec, Shell, Saudi ARAMCO, and SABIC, on the development of commercial catalysts.

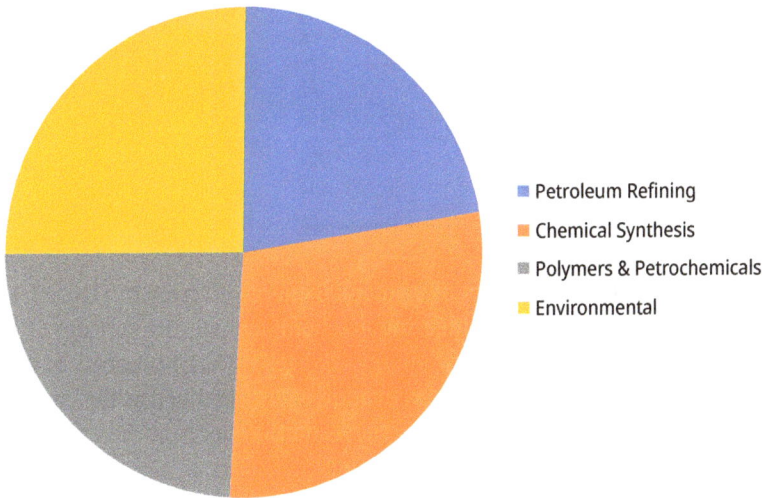

- Petroleum Refining
- Chemical Synthesis
- Polymers & Petrochemicals
- Environmental

Figure 1.3: Global catalyst market share (%), by application, in 2019.
Source: http://www.grandviewresearch.com.

1.2 General characteristics of solid catalysts

This section briefly considers some of the key characteristics in heterogeneous catalysis and introduces the terminology for discussing solid catalysts.

---

2 R&D is a shorthand notation for research and development activities undertaken by companies or government institutions to develop new processes or products or improve existing processes or products.

1.2.1 Activation energy

A catalyst provides a new, alternative, pathway for the conversion of reacting molecules to product molecules. The overall activation energy for a catalyzed reaction is substantially lower than that of the same reaction without a catalyst. Table 1.1 presents examples of the reduction in activation energy for some heterogeneous and homogeneous reactions.

Table 1.1: Activation energy for catalyzed and uncatalyzed reactions [1].

Reaction	Catalyst	E (kJ/mol)
$2\ HI \rightarrow H_2 + I_2$	$-$	184
	Au	105
	Pt	59
$2\ N_2O \rightarrow 2\ N_2 + O_2$	$-$	245
	Au	121
	Pt	134
Pyrolysis of $(C_2H_5)_2O$	$-$	224
	I_2 vapor	144

This situation is demonstrated in the form of a potential energy diagram in Figure 1.4. Students of chemistry and chemical engineering are familiar with the concept of a potential energy diagram: It represents how energy changes with the progress of the reaction. It shows the energy of reactants, the activation energy (defined as the energy required by the reactants to cross the energy barrier), and the energy of the products. Is the reaction in Figure 1.4 endothermic or exothermic?

The strong effect of this reduction in the activation energy can be demonstrated by considering the hydrogenation of ethylene in the absence and presence of a CuO–MgO-based catalyst. The reaction rate equations are given by the following expressions, where P is the pressure [2]:

$$r_{uncat} = 10^{27} \exp(-43,000/RT)P_{H_2}$$

$$r_{cat} = 2 \times 10^{27} \exp(-13,000/RT)P_{H_2}$$

The student is invited to calculate the ratio of the two rates (r_{cat}/r_{uncat}) at, say, 600 K, to see the huge impact of about 2/3 drop in the activation energy.

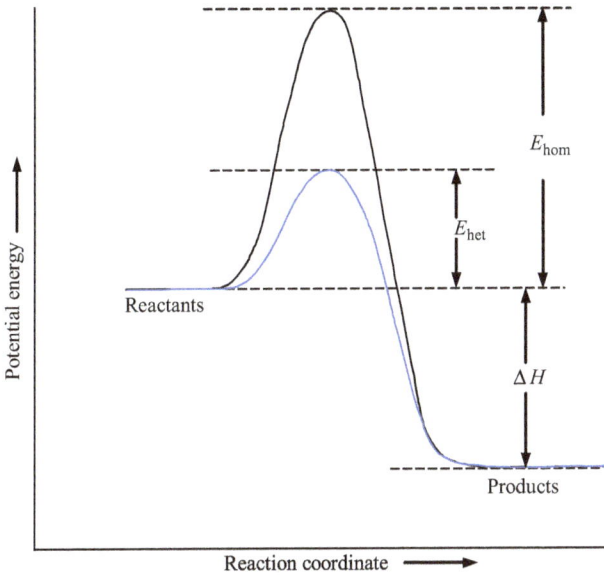

Figure 1.4: The effect of catalysis on the activation energy.

1.2.2 Classification by functionality

Catalysis is essentially a chemical phenomenon. The ability of a substance to act as a catalyst for a specified reaction system depends on its chemical nature. For catalysis to occur, there must be a chemical interaction between the catalyst and the reactant-product system.

However, such an interaction, as we mentioned earlier, will not change the chemical nature of the catalyst except at the surface where molecules are *adsorbed* (i.e., attached) during the reaction sequence.

As we shall see in the coming chapters, it is essential to understand the **adsorption** of molecules at solid surfaces before we try to acquire a deeper knowledge of heterogeneous catalysis.

In simple terms, the *functionality* of the catalyst refers to its ability to efficiently perform a specific action. Experience has shown that the functionality depends on the chemical nature of the catalyst. Table 1.2 shows a *broad* classification of some solid catalysts depending on their catalytic abilities.

It is worth noting here that a semiconducting material is a compound whose electrical conductivity is intermediate between that of a metal (e.g., tungsten or platinum) and an insulator (e.g., rubber or glass).

Table 1.2: Solid catalysts classified by functionality.

Class	Examples	Functionality
Metals	Fe, Ni, Pd, Pt, Ag, Cu	Hydrogenation, dehydrogenation, oxidation
Semiconducting metal oxides and sulfides	NiO, ZnO, MnO_2, Cr_2O_3, Bi_2O_3–MoO_3, WS_2	Oxidation, dehydrogenation, desulfurization
Acidic solids	SiO_2–Al_2O_3, zeolites	Dehydration, polymerization, isomerization, cracking, alkylation

1.2.3 Selectivity of solid catalysts

While the **activity** of a catalyst refers to the rate at which it causes the reaction to proceed to chemical equilibrium, another important benefit of solid catalysts is the selectivity. The *selectivity* of a catalyst refers to its ability to form one or more of the desired products.

Most of us probably remember from our background in homogenous reactor design that the selectivity depends on factors such as the operating pressure and temperature, reactor type, and conversion level. Now, we can add the catalyst.

Let us consider the decomposition of ethanol, as shown below, to illustrate the meaning of catalyst selectivity. In case (1), a *copper-based* catalyst is used where acetaldehyde and hydrogen result from the decomposition (note that Cu adsorbs H_2). In case (2), however, the use of an *alumina-based* oxide catalyst leads to the adsorption of H_2O:

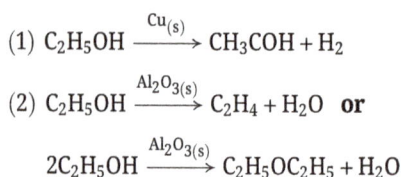

$$(1)\ C_2H_5OH \xrightarrow{Cu_{(s)}} CH_3COH + H_2$$

$$(2)\ C_2H_5OH \xrightarrow{Al_2O_{3(s)}} C_2H_4 + H_2O \ \ \textbf{or}$$

$$2C_2H_5OH \xrightarrow{Al_2O_{3(s)}} C_2H_5OC_2H_5 + H_2O$$

Therefore, we can conclude that the functionality of the catalyst also determines its selectivity.

A catalyst may be useful for its activity, selectivity, or both. In the case of multiple reactions (e.g., parallel, series, or parallel–series), the selectivity is generally more important. It should also be mentioned here that *most* industrial reactions are multiple in nature.

1.2.4 The catalytic site

An important concept in heterogeneous catalysis is the *catalytic site*. This concept was introduced by H.S. Taylor in 1925 [3]. Catalytic sites are specific locations on the solid surface (e.g., clusters of atoms located on lattice defects, edges, and corners) where the catalytic reaction predominantly takes place.[3] These locations have also been called *active centers*. At the molecular level, many of these sites or active centers may be present in a small area of the catalyst surface.

1.2.5 Composition and preparation of solid catalysts

Solid catalysts are generally composed of three components: the **active agents** (metals, metal oxides, etc.); the **support** materials (often called the **carrier**); and the **promoter** materials.

Most industrial catalysts are *supported*, but there are cases in which no support is used, and such solids are called *unsupported* (e.g., the active agent is in the form of wires or gauzes). An example of catalytic gauze disks is shown in Figure 1.5.

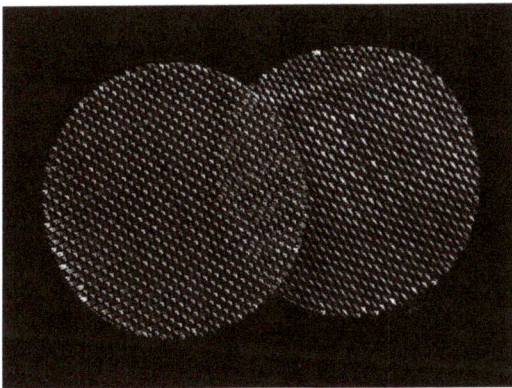

Figure 1.5: Image of Pt gauze disks (https://www.elementalmicroanalysis.com).

In supported catalysts, the catalytic agent (e.g., metals or metal oxides) is dispersed on another substance (the support or carrier), in the form of small crystals (10–100 Å).

The active agent(s) typically forms 10–20% by weight of the total catalyst. An example of this configuration is shown in Figure 1.6.

---

3 This is a rather simplistic view of the active site. I encourage the reader to look up the paper by Zambelli et al. [4] and the recent work by Vogt and Weckhuysen [5]. The latter reference contains a historical review and an extensive analysis of the concept of active sites.

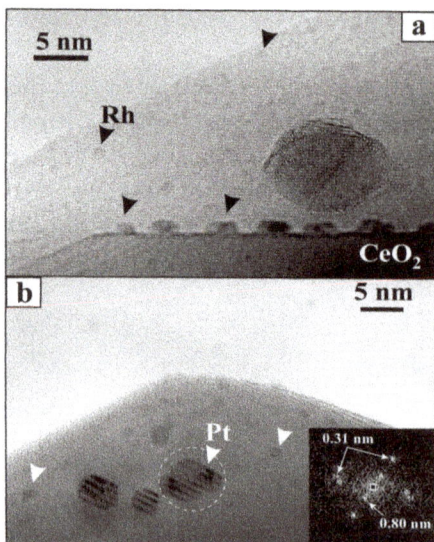

Figure 1.6: High-resolution electron microscopy (HREM) images of a supported Rh–Pt–CeO$_2$/Al$_2$O$_3$ catalyst [6].

The support solids can have a variety of shapes; for example, small cylindrical pellets, tablets, spheres, extrudates, or honeycomb monoliths. Some of those shapes and others are shown in Figure 1.1.

Table 1.3 lists typical industrial support materials. The support may be chemically active or inert. In any case, the following characteristics are normally desirable in the support:

- High *porosity* (to provide large surface area).
- Thermal and chemical *stability* (to resist typically severe operating conditions in industrial reactors).
- Mechanical *strength* (to allow the packing of large catalyst amounts in industrial packed-bed reactors or the recycling of large amounts in other cases).

Table 1.3: Typical industrial support materials [7].

Support	Specific area (m^2/g)
Activated carbon	500–1,500
Silica gel	200–800
Kieselguhr (naturally occurring silica)	15–20
Powdered silica glass (SiO$_2$)	0.1–0.6
Activated alumina (Al$_2$O$_3$)	150–400
Silica–alumina (SiO$_2$–Al$_2$O$_3$)	200–700
Zeolites and natural clays	500–700

For these reasons, the support is typically a **porous** material characterized by a large surface area, because the catalyst activity is generally proportional to the concentration of active sites. Table 1.3 also lists typical surface areas.

In addition to the active agent(s) and support, most catalyst formulations include promoters. The **promoter** is a substance that may not be catalytically active itself, but it is added, during catalyst synthesis or during the reaction, to allow the active agent(s) to function to its maximum capacity. In other words, the promoter improves the effectiveness of the catalyst.

Promoters are classified as either physical or chemical. **Physical** promoters are added for the purpose of maintaining the physical and structural integrity of the support or active agent(s). **Chemical** promoters on the other hand are added to increase the intrinsic activity of the catalyst. Let us consider some examples.

Example 1.1:

$$3\,H_2 + N_2 \xrightarrow[+H_2O_3]{Fe_3O_{4(s)}} 2\,NH_3$$

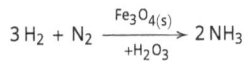

As we have seen earlier, the iron-oxide catalyst is magnetite. In addition, a small amount of H_2O_3 is included as a physical promoter to prevent "sintering" of the Fe crystallites, that is, to prevent the catalyst (under severe operating conditions) from changing into a nonporous mass with a much smaller surface area.

Example 1.2:

$$3\,H_2 + N_2 \xrightarrow[+K_2O]{Fe_3O_{4(s)}} 2\,NH_3$$

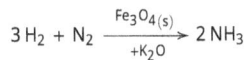

Here K_2O is a chemical promoter that is used to increase the activity of Fe crystallites.

Example 1.3:

$$C_2H_4 + 1/2\,O_2 \xrightarrow[\text{in feed}]{C_2H_4Cl_{2(ppm)}} C_2H_4O$$

$$C_2H_4 + 3\,O_2 \rightarrow 2\,CO_2 + 2\,H_2O$$

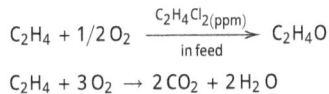

In this case, ethylene dichloride is added in parts per million (ppm) to the feed as a chemical promoter that inhibits the second parallel reaction where the complete oxidation of ethylene is undesirable.

We should not end this section without looking at how such complex catalyst particles are made in practice. As we might expect, **catalyst preparation** is sophisticated and contains many details that must be clear to the catalyst manufacturer. Certainly, the large number of details in the preparation stage affects the final properties of the catalyst, especially the catalytic efficiency and selectivity.

The general steps for preparation of the catalyst include selection of the ingredients, mixing of the ingredients, mixing with a binder or forming agent, shape formation, calcination, and activation.

Some of the commonly used techniques for the preparation of catalysts are *precipitation, sol-gel, ion exchange*, and *impregnation*.

Further examination of solid catalyst preparation techniques is given in the textbook website at www.degruyter.com.

1.2.6 Naming of solid catalysts

According to the preceding section, it is now clear that most catalysts are composed of many substances. An example of the complex composition of solid catalysts is given in Table 1.4, where the composition of an industrial naphtha hydrodesulfurization catalyst is presented.

Table 1.4: The chemical composition of a UOP hydrotreating catalyst[a].

Substance	Formula	Weight %
Molybdenum oxide	MoO_3	15.0
Cobalt oxide	CoO	4.0
Aluminum oxide	Al_2O_3	74.7
Sodium oxide	Na_2O	0.08
Potassium oxide	K_2O	0.09
Calcium oxide	CaO	0.02
Magnesium oxide	MgO	0.07
Silicon oxide	SiO_2	0.15
Iron oxide	Fe_2O_3	<0.02
Sulfate	SO_4	0.07
Vanadium oxide	V_2O_5	<0.02
Nickel oxide	NiO	0.03
Phosphorus oxide	P_2O_5	0.07

[a]Courtesy of the Research Institute, KFUPM.

How is such a multi-substance catalyst named? Well, it is routine to name the catalyst by listing only the *active element(s)* and *support* (the carrier), without specifying the form in which the active element may exist either in the catalyst as manufactured or under reaction conditions (for example, many catalysts are usually supplied as metal oxides but converted to sulfides before use).

It then follows that the hydrodesulfurization catalyst listed in Table 1.4 is named "CoMo/Al_2O_3," or in the everyday language of process engineers, "Cobalt Moly on Alumina." This name implies that the active agents are based on cobalt and molybdenum, and the support is alumina.

1.3 The catalytic sequence

The catalytic reaction process involves the *reactor* in which the catalytic *reaction* takes place. In the reactor, we speak of the *global* catalytic reaction rate, whereas we speak of the *intrinsic* reaction rate at the surface of the catalyst.

The catalytic packed-bed reactor (also called the fixed-bed reactor) is commonly used in industry. A schematic of a typical PBR is shown in Figure 1.7a. The design, performance, and reliability of the PBR are of great industrial importance.

The subject of reactor performance is addressed later in this book, but we first need to understand what happens *around* and *within* a catalyst particle. In other words, we need to understand what happens around and within a "point" in the reactor. Figure 1.7b demonstrates the concept of a point in the reactor.

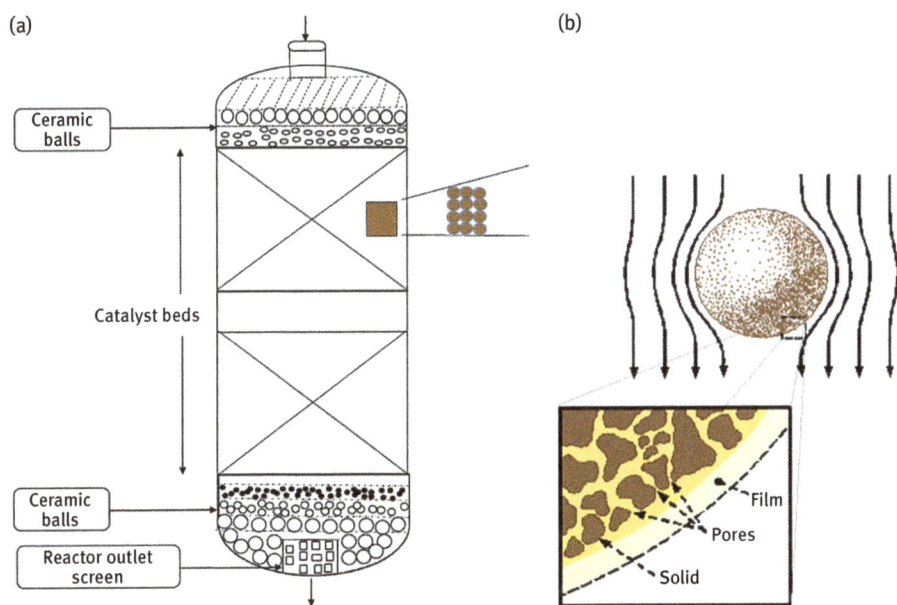

Figure 1.7: (a) Schematic illustration of an industrial packed-bed reactor and the individual catalyst particles. (b) Concept of a "point" in a catalytic reactor (adapted from [8]).

The catalytic *reaction sequence* is defined as the steps involved in the reaction at a point in the reactor. The sequence is composed of several steps that describe transport of reactants from the *bulk* fluid phase, the reaction on the catalyst surface, and the transport of products back into the bulk fluid phase. For the general catalytic reaction,

$$A + B + \cdots \Leftrightarrow R + S + \cdots$$

several steps are involved:
(1) *Transport* of reactants A, B, . . . from the bulk fluid phase within the reactor to the external surface of the catalyst.
(2) *Transport* of reactants to the internal surface of the pores of the catalyst.
(3) *Adsorption* of the reactants onto the catalytic sites.
(4) Chemical *reaction* between adsorbed species.
(5) *Desorption* of products R, S, . . . from the catalytic sites.
(6) *Transport* of products from inside the pores back to the surface of the catalyst.
(7) *Transport* of products from the catalyst surface to the bulk fluid phase.

These seven steps are shown in detail in Figure 1.8. Note that we shall adopt the following terminology here and in other chapters: in steps (1) and (7), C_{Ab} and C_{Rb} are concentrations of reactant A and product R in the bulk fluid; C_{As} and C_{Rs} refer to concentrations at the external surface of the catalyst particle; in steps (2) and (4), AX and RX refer to the adsorbed reactant and the product; C_A and C_R refer to concentrations within the pore.

1- reactants external
   mass transfer
2- reactants internal
   diffusion
3- reactants adsorption
4- surface reaction
5- products desorption
6- products internal
   diffusion
7- products external
   mass transfer

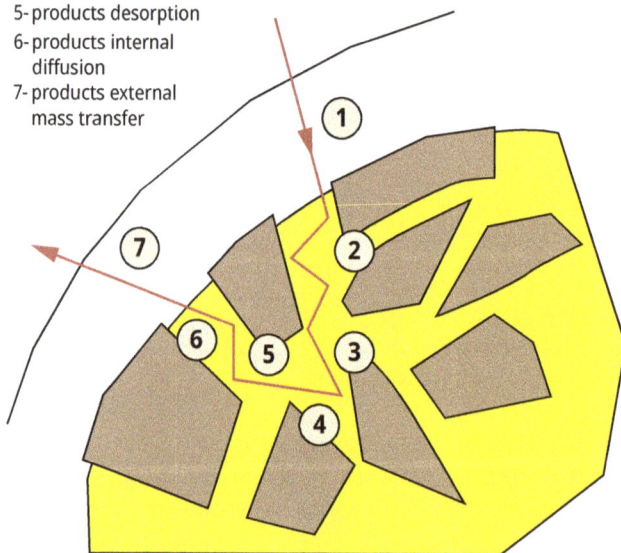

Figure 1.8: Steps involved for reactions on porous solid catalysts (adapted from [9]).

Let us carefully consider these steps. We see that, for solid-catalyzed reactions, not only do we have to worry about the chemical reaction itself (with associated heat effects) but we also need to consider the physical transport mechanisms (convection, diffusion, and conduction).

It is common in the catalytic reaction engineering literature to refer to steps (1) and (7) as **"external"** transport steps, that is, transport of species *between* two phases or regions (the bulk fluid and solid).

Likewise, it is common to call steps (2) and (6) **"internal"** transport steps,[4] that is, transport of species *within* a phase or region (the solid).

In our subsequent discussions, steps (3) – (5) are called the catalytic reaction cycle. An example of a catalytic reaction cycle, involving the formation of CO_2 by the oxidation of CO, is shown in Figure 1.9

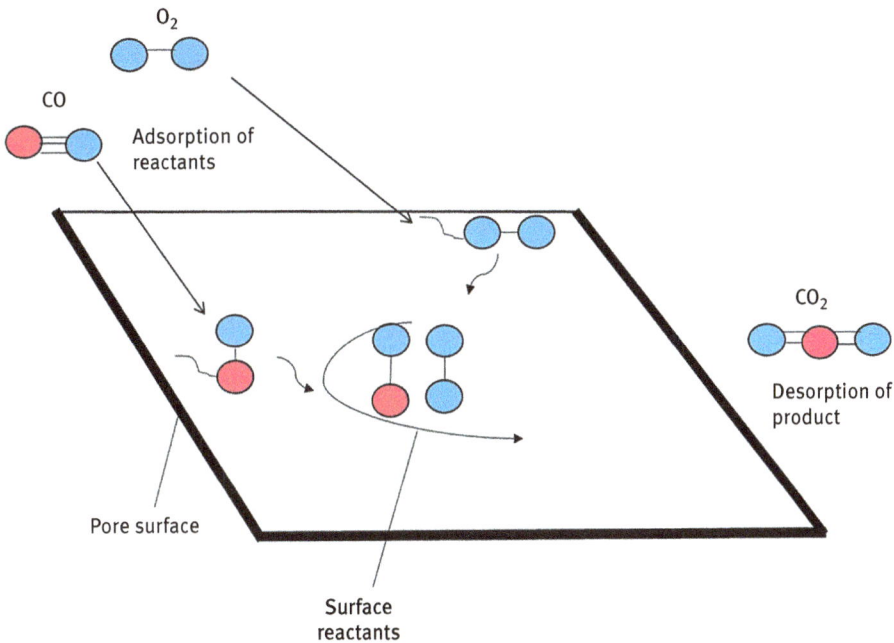

Figure 1.9: The catalytic reaction cycle in the reaction of adsorbed CO and O_2 on the pore surface.

Most of the first part of this book deals with these individual steps in detail. Thus, in Chapter 2 we shall study the subject of adsorption; Chapter 4 discusses surface reactions; and in Chapter 5 we shall study internal and external transport. It is essential to

---

4 In the literature, other names for external and internal transport can be found, such as *inter-phase* and *intra-phase transport*, respectively.

treat these topics in depth before we take up the question of catalytic reactor design in Chapter 7.

References

[1]   Bond G. C. "Heterogeneous Catalysis: Principles & Applications," Second Ed., Oxford University Press, Oxford, 1987.
[2]   Smith J. M. "Chemical Engineering Kinetics," Third Ed., McGraw-Hill, New York, 1981.
[3]   Laidler K. J. "Chemical Kinetics," Third Ed., Harper & Row, New York, 1987.
[4]   Zambelli T., Wintterlin J., Trost J., and Ertl G. "Identification of the "Active Sites" of a Surface-Catalyzed Reaction," Science, 273, 1688 (1996).
[5]   Vogt C., Weckhuysen B. M. "The Concept of Active Site in Heterogenous Catalysis," Nat. Rev., 6, 89 (2022).
[6]   Martiniz L. M., Laguna O. H., Lopez-Cartez C., and Centeno M. A. "Synthesis and Characterization of Rh/MnO_2-CeO_2/Al_2O_3 for Co-PrO$_x$ Reaction," Molec. Catal., 440, 9 (2017).
[7]   Clausen III C. A. and Mattson G. "Principles of Industrial Chemistry," Wiley, New York, (1978).
[8]   Fulton J. W. "Building the Mathematical Model of the Catalyst and Reactor," Chemical Engineering, Part I, 118, February, (1986).
[9]   Masi M. "Chemical Reactors, Chapter 6.3, Vol. 5," Encyclopedia of Hydrocarbons, Treccani Reti, Rome, (2008).

2 Adsorption

First this. Then that.
Michael Robbins

At the end of the previous introductory chapter, we discussed how the global rate of a solid-catalyzed reaction depends on several steps, including mass transfer, adsorption, and surface reactions.

Beginning in this chapter, we shall take up these steps in more detail. Here we shall attempt to answer questions related, among other things, to the following:
- the surface concentration of adsorbed species,
- how rapidly reactants will establish themselves on active sites, and
- the relationship between surface concentrations and the measurable bulk-phase concentrations.

2.1 Types of adsorption

If we use a microscope to look at even the most polished surfaces we will probably find that such surfaces are *not smooth*. In general, surfaces at that level are irregular with "fissures," "peaks," and other topographical discontinuities, as shown in Figure 2.1. These regions of irregularity are particularly subject to residual force fields [1]. At these locations, surface atoms of the solid may attract other atoms or molecules in the surrounding fluid phase. Similarly, the surfaces of pure crystals, see Figure 2.2 for an example, have nonuniform force fields because of the atomic structure in the crystal. Such surfaces also contain active sites where adsorption is enhanced [1].

Figure 2.1: Scanning electron micrograph (SEM) of a commercial metal oxide/SiO_2–Al_2O_3 catalyst (courtesy of Research Institute, KFUPM).

https://doi.org/10.1515/9783111032511-002

Figure 2.2: Scanning electron micrograph (SEM) of a high-silica zeolite catalyst (courtesy of Research Institute, KFUPM).

Under suitable temperature and pressure conditions, a gas will adsorb upon a solid, finally covering its surface entirely. The term "suitable" conditions typically means very low temperatures and relatively high pressures.

Two forms of adsorption may occur: **physical** adsorption (also called physisorption) and **chemical** adsorption (commonly called chemisorption).

We distinguish between the two types based on the *strength of the binding forces*, that is, forces attracting fluid molecules (the *adsorbate*) to the solid surface (the *adsorbent*).

In physical adsorption, the binding forces are relatively weak. These forces are proportional to van der Waals forces (recall that those forces are the ones considered in the van der Waals equation of state). Physical adsorption is somewhat similar to condensation (or liquefaction) and is *non-specific*, that is, under suitable conditions it would occur on all surfaces.

In chemical adsorption, however, the bonding between molecules of *certain* fluids and the surfaces of *specific* solids is quite strong. Unlike the first type of adsorption, the binding forces in chemical adsorption are like those characterizing chemical compound formation, that is, the so-called valence forces.

Compared to physical adsorption, this type is *specific*, meaning that not all solids will chemisorb a given gas or group of gases. In catalysis, the chemical adsorption of *at least one reactant* is a prerequisite in the reaction sequence.

The **specificity** of adsorption is an important property that provides us with a means of estimating the *total surface area* of the solid versus the area occupied by active sites, which we may call the *catalytic area*.

2.2 Comparison between physical and chemical adsorption

Because of the importance of adsorption in catalytic reactions, we shall now discuss several properties of the two types of adsorption.

2.2.1 Surface coverage

Physical adsorption is not highly dependent on surface irregularities but is usually proportional to the amount of surface. However, the extent of adsorption is not limited to a monomolecular layer on the surface as shown in Figure 2.3, but multiple layers can form.

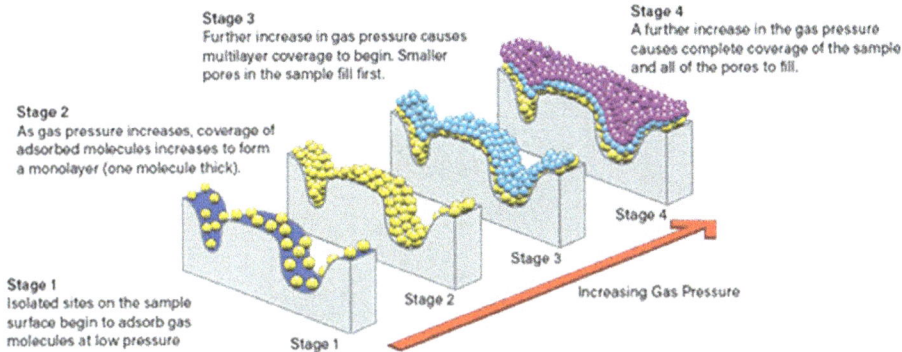

Stage 3
Further increase in gas pressure causes multilayer coverage to begin. Smaller pores in the sample fill first.

Stage 4
A further increase in the gas pressure causes complete coverage of the sample and all of the pores to fill.

Stage 2
As gas pressure increases, coverage of adsorbed molecules increases to form a monolayer (one molecule thick).

Stage 1
Isolated sites on the sample surface begin to adsorb gas molecules at low pressure

Stage 4

Stage 3

Stage 2

Stage 1

Increasing Gas Pressure

Figure 2.3: Multilayer form of molecule–solid attachment [2].

By contrast, in chemisorption, *only* a monolayer forms (as in Stage 1) on the surface. This limitation results from the valence forces holding molecules on the surface which weaken rapidly with distance [1].

2.2.2 Heat of adsorption

Heat is usually released during adsorption; that is, adsorption is exothermic. The heat of adsorption (ΔH_{ads}) is dependent on the type.

In physical adsorption, ΔH_{ads} is of the same order of magnitude as the heat of condensation ΔH_{cond} (i.e., in the range ~ 0.5–5 kcal/mol). By contrast, ΔH_{ads} in chemisorption is of the same order of magnitude as the heat of reaction ΔH_R (i.e., in the range ~ 5–100 kcal/mol).

Because of the relatively high ΔH_{ads}, the energy of chemisorbed molecules can be substantially different from that of the molecules alone. Hence, the activation energy (E) for reactions involving chemisorbed molecules can be much smaller than the activation energy for reactions involving only, say, gas-phase molecules. On this basis, chemisorption offers an explanation for the catalytic effect of solid surfaces.

2.2.3 Catalytic cycle

Let us try to clarify the last point in the previous section by considering again the reaction pathways of a simple exothermic reaction as shown in Figure 2.4.

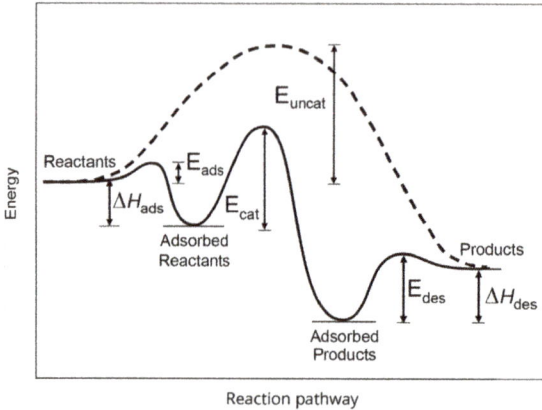

Reaction pathway

Figure 2.4: Energy changes associated with individual steps in a simple exothermic reaction (- - - - - uncatalyzed and —— catalytic).

What we have called the catalytic cycle is composed of three steps: chemisorption, formation and breakup of an activated complex, and desorption. As shown in Figure 2.4, each step has its activation energy. These are defined as follows:

E_{ads}: activation energy for adsorption of reactants

E_{cat}: activation energy for formation of activated complex

E_{des}: activation energy for desorption of products

ΔH_{ads}: heat of adsorption of reactants, taken to be *exothermic*

ΔH_{des}: heat of desorption of products, taken to be *endothermic*

ΔH_{R}: overall enthalpy change upon reaction; the same for both reactions

At this point, some of us might wonder how the activation energy of a catalytic reaction is determined.

Recall that in homogeneous reactions, we usually calculate E from the slope of an Arrhenius diagram (see [3] for examples). In catalytic reactions, experimental rate data can be collected, and a so-called *apparent* activation energy can be calculated from the slope of an Arrhenius diagram. How then do we get the *true* activation energy for the surface reaction?

In general, this would require knowledge (or assumption) of the mechanism of the surface reaction, identification of the rate-limiting step, ΔH_{ads} and ΔH_{des}. We will re-examine this issue in Chapter 4.

2.2.4 Rate of adsorption

In physical adsorption, equilibrium between the solid surface and fluid molecules is (usually) rapidly attained and easily reversible because the energy requirements are small. Recall that typically $E_{ads} \sim$ 0–1 kcal/mol.

By contrast, two kinds of chemisorption are encountered: *activated* (E_{ads} = finite) and *nonactivated* ($E_{ads} \sim$ 0 and therefore this kind occurs rapidly). The first kind is more frequent whereas the second is less frequent. It is often found that for a given gas–solid system, the initial chemisorption is nonactivated, while later stages of the process are slow and temperature-dependent.

The *amount* of physical adsorption decreases rapidly as the temperature increases and is generally very small above the critical temperature, T_c, of the adsorbed gas. The amount due to activated chemisorption increases through a maximum at moderate temperatures and drops at higher temperatures as the rate of desorption exceeds that of adsorption. Figures 2.5a and b illustrate the effect of temperature on the amount adsorbed.

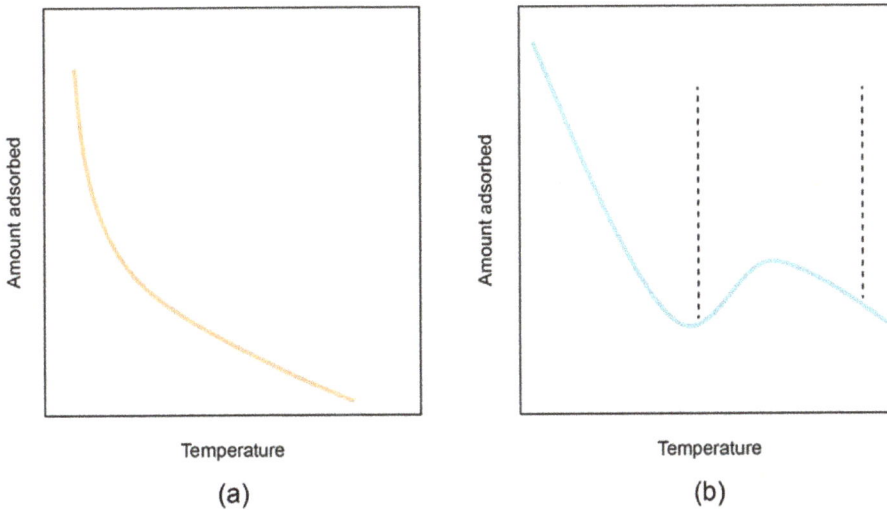

Figure 2.5: Effect of temperature on the amount adsorbed: (a) physisorption and (b) chemisorption. The area between the dashed lines represents the range of potential catalytic activity.

We end this section by summarizing the differences between physical and chemical adsorption in Table 2.1.

Table 2.1: Comparison between physical and chemical adsorption.

Parameter	Physical	Chemical
Adsorbent (the solid)	All solids	Some solids
Adsorbate (the fluid)	All gases, $T < T_c$	Some chemically reactive gases
Temperature range	Low temperature	Possible over a wide range and generally high temperatures involved
Heat of adsorption	Low, $\sim \Delta H_{cond}$	High, $\sim \Delta H_R$
Activation energy and rate	Low E and very rapid	Nonactivated: low E; activated: high E
Coverage	Multilayer possible	Monolayer
Reversibility	Highly reversible	Often irreversible

2.3 Adsorption isotherms

We turn our attention now to the *quantitative* description of the relation between the amount of a substance adsorbed and its partial pressure in the gas phase (in the case of liquids, the concentration replaces the pressure).

We require this description in order to (a) describe the extent and strength of adsorption, (b) establish the kinetics of catalytic reactions, and (c) estimate the surface area of the solid.

The term "adsorption isotherm" is more precisely defined as "the relation, at constant temperature, between the quantity of gas adsorbed and the pressure with which it is in equilibrium."

Figure 2.6 schematically shows what is involved in determining the quantity adsorbed. The amount of gas adsorbed is usually followed either gravimetrically or volumetrically. In the *gravimetrical* method, the solid sample is weighed using a microbalance during the experiment. In the *volumetric* method, the gas consumption is followed by measuring the difference in the rates of flow of gas into and out of the system.

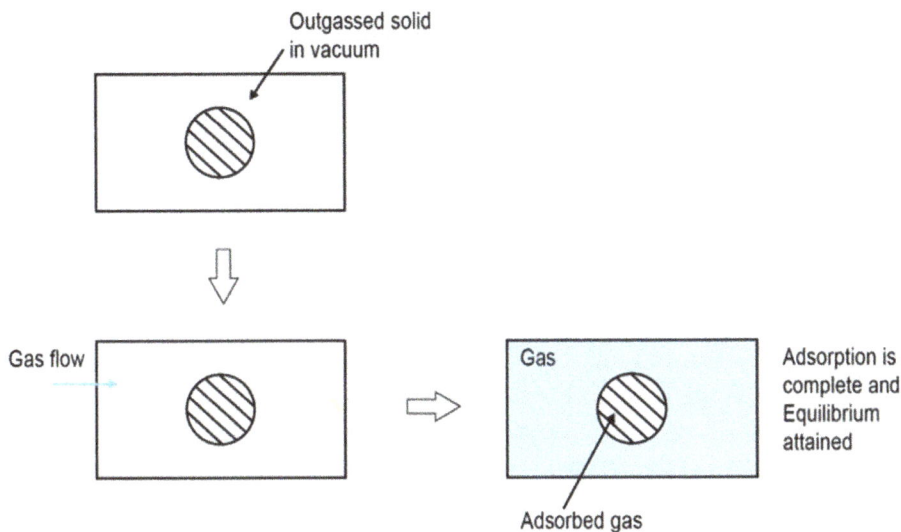

Figure 2.6: Schematic diagram showing steps involved in experimental determination of the amount of gas adsorbed.

2.3.1 Langmuir isotherm

Over the years several adsorption isotherms were derived. The first attempt to relate the amount adsorbed to its concentration in the gas phase was done by Langmuir[1] [4, 5]. Briefly, the assumptions involved in deriving the Langmuir isotherm are:

(1) Each unit area of surface consists of n unit sites, each of which can adsorb one, but only one, molecule of the gas.
(2) All sites are energetically equivalent.
(3) The adsorbed molecules do not interact.
(4) The maximum possible adsorption corresponds to a monolayer.

We shall now derive the Langmuir isotherm. Suppose that, after equilibrium is established, a fraction θ of the surface is *covered* by the adsorbed molecules; it follows then that the fraction of surface *uncovered* is $(1 - \theta)$.

It is clear that the rate of adsorption, r_a, will be proportional to the concentration of molecules in the gas phase (or in solution) and will also depend on the fraction of surface uncovered, since adsorption can only occur when molecules strike the bare surface. Therefore, we can express the rate of adsorption as follows:

---

1 Langmuir was awarded the Nobel Prize in chemistry in 1932 for his work on surface science.

$$r_a = k_a C_A (1 - \theta) \tag{2.1}$$

where k_a is the adsorption constant. The rate of desorption, r_d, is proportional to the number of molecules attached to the surface and this, in turn, is proportional to the fraction of surface covered. Therefore, we can write:

$$r_d = k_d \theta \tag{2.2}$$

where k_d is the desorption constant.

At equilibrium, the rates of adsorption and desorption are equal. Thus, by using eqs. (2.1) and (2.2), and solving for θ, we obtain:

$$\theta = \frac{K C_A}{1 + K C_A} \tag{2.3}$$

where K is defined as the *adsorption equilibrium constant*, that is, $K = (k_a/k_d)$. Note that for gaseous systems, the concentration terms can be replaced by partial pressure terms.

An alternative method by which we can arrive at the Langmuir isotherm is used in reaction kinetics. Here the adsorption equilibrium can be represented as follows:

$$A + X \Leftrightarrow AX$$

where A is the adsorbing molecule, X is the *adsorption site*, and AX represents the adsorbed complex.

As we usually do when we look at the kinetics of chemical reactions, we can define an equilibrium constant:

$$K = \left[\frac{C_{AX}}{C_A C_X} \right]_{eq} \tag{2.4}$$

Now, C_A can be replaced by P_A, the pressure of A at equilibrium; C_X can be replaced by $n (1 - \theta)$, which is the concentration of vacant sites; and C_{AX} by $n\theta$, which equals the concentration of occupied sites. Thus,

$$K = \frac{n\theta}{P_A n (1 - \theta)} = \frac{\theta}{P_A (1 - \theta)}$$

from which we obtain,

$$\theta = \frac{K P_A}{1 + K P_A} \tag{2.5}$$

Note that K, the adsorption constant of A, reflects the strength of adsorption. When K is large, we say that A is *strongly adsorbed* (or the surface is well covered), and when K is small, we say that A is *weakly adsorbed* (or the surface is sparsely covered). In other words, the larger the value of K, the greater the surface coverage at a given equilibrium pressure.

Also note that since adsorption is almost always exothermic, then application of the *Le Chatelier's principle* shows that K decreases as temperature increases. The dependence of K on temperature is given by the van't Hoff equation: $K = C \exp\left(-\Delta H^0_{ads}/RT\right)$. (At this point, it is useful to remind yourself of the statement of this principle by consulting standard thermodynamics textbooks.) The effect of P_A and K is shown in Figure 2.7a and b.

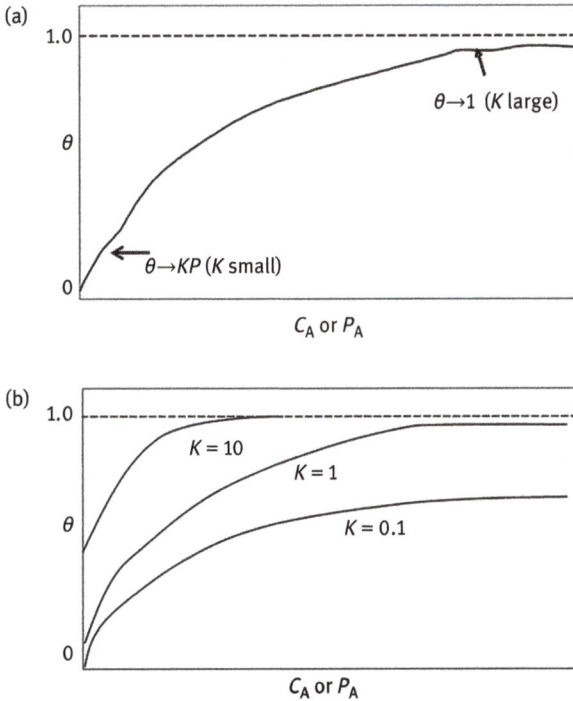

Figure 2.7: Behavior of the Langmuir adsorption isotherm at (a) asymptotic limits of K and (b) finite values of K.

2.3.2 Competitive adsorption

Let us now see what happens when two substances, A and B, compete to adsorb on the same surface, that is, what is the form of the isotherm in this case?

We start by defining θ_j as the fraction of surface covered by molecules of species j. Thus, $(1 - \Sigma \theta_j)$ = fraction bare. In the case at hand, the fraction bare is $(1 - \theta_A - \theta_B)$.

If both species, A and B, are adsorbed, then as we have done in the preceding section, we can write expressions for the rates of adsorption and desorption as follows:

$$r^A_a = k^A_a C_A (1 - \theta_A - \theta_B) \tag{2.6}$$

$$r_a^B = k_a^B C_B (1 - \theta_A - \theta_B) \tag{2.7}$$

$$r_d^A = k_d^A \theta_A \tag{2.8}$$

$$r_d^B = k_d^B \theta_B \tag{2.9}$$

At equilibrium, $r_a^A = r_d^A$ and $r_a^B = r_d^B$, from which we obtain:

$$\frac{\theta_A}{1 - \theta_A - \theta_B} = K_A C_A \tag{2.10}$$

$$\frac{\theta_B}{1 - \theta_A - \theta_B} = K_B C_B \tag{2.11}$$

where $K_j = \left(k_a^j / k_d^j\right)$ is the adsorption coefficient for species j. Solving eqs. (2.10) and (2.11) simultaneously, we obtain:

$$\theta_A = \frac{K_A C_A}{1 + K_A C_A + K_B C_B} \tag{2.12}$$

$$\theta_B = \frac{K_B C_B}{1 + K_A C_A + K_B C_B} \tag{2.13}$$

In terms of partial pressures, these two equations become

$$\theta_A = \frac{K_A P_A}{1 + K_A P_A + K_B P_B} \tag{2.14}$$

$$\theta_B = \frac{K_B P_B}{1 + K_A P_A + K_B P_B} \tag{2.15}$$

It is instructive to check what happens to these equations when $P_B = 0$ or $K_B = 0$.

It is important to mention that an implied assumption in the above analysis is that species A and B are adsorbed *without* molecular dissociation on the surface. That interesting case is considered in Problem 6 at the end of the textbook.

2.3.3 Determination of surface area

According to the Langmuir isotherm, the fraction of sites occupied, θ, is proportional to the volume of gas adsorbed, V, which is less than or equal to the volume of gas adsorbed at complete **monolayer** coverage, V_m. Therefore, we can define θ as equal to the volume of gas adsorbed to that which would be adsorbed in a monolayer, or

$$\theta = \frac{V}{V_m} \tag{2.16}$$

Both volumes are typically reported at standard conditions (STP) or at fixed temperature and pressure conditions.

Combining eqs. (2.5) and (2.16), we obtain the following relation between the pressure and the amount adsorbed:

$$V = \frac{V_m K P_A}{1 + K P_A} \tag{2.17}$$

Note that at low pressures, V changes linearly with pressure; and at high pressures, $V \rightarrow V_m$ as expected. Equation (2.17) can be rearranged to be more suitable for the analysis of experimental data, which is the following:

$$\frac{P_A}{V} = \frac{1}{V_m K} + \frac{P_A}{V_m} \tag{2.18}$$

If the Langmuir isotherm is obeyed, a plot of (P_A/V) versus P_A should be linear and the slope is $(1/V_m)$, as shown in Figure 2.8.[2]

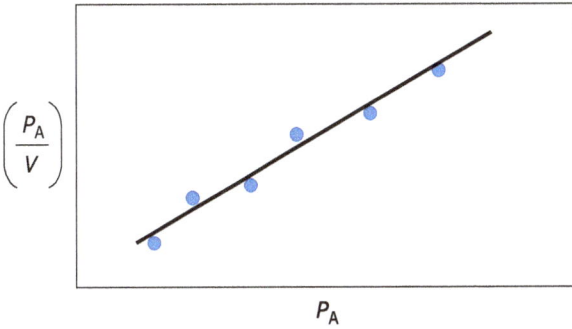

Figure 2.8: Schematic representation of eq. (2.18); the dots represent experimental data.

Once V_m is determined, it can be converted to the number of molecules adsorbed by dividing by the molar volume at reference conditions, V_{ref}, and multiplying by Avogadro's number (N_o), thus

$$\text{Number of molecules adsorbed} = \frac{V_m}{V_{ref}} N_o$$

Usually V_m is recorded at STP, that is at $T = 273$ K and $P = 1$ atm; and the value of $V_{ref} = 22,400$ cm^3/mol at these conditions as we know. Next, we can multiply this number by the area covered by an adsorbed molecule, a, and divide by the weight of the solid sample employed in the experiments in question, to obtain the *specific surface area*:

---

2 Butt [12] recommends an alternative linear form: $(1/V) = (1/V_m) + (1/KV_m)(1/P_A)$.

$$S_g = \frac{\left[\frac{V_m}{V_{ref}} N_o\right] \alpha}{\text{weight of solid}}$$

(2.19)

Smith [1] suggests that we use the following formula to compute α, which is based on the projected area of a molecule when the molecules are arranged in a closed 2-dimensional packing:

$$\alpha = 1.09 \left[\frac{M}{N_o \rho}\right]^{2/3}$$

(2.20)

where M is the molecular weight of the adsorbate, and ρ is the density of the adsorbate taken as that of pure liquid at the temperature of the experiment. For example, $\rho = 0.808$ g/cm^3 for N_2 at −195.8 °C. Therefore, using eq. (2.20) we can compute $\alpha = 16.2$ (Å)^2 or 16.2×10^{-16} cm^2. We will say more about the surface area in the next section.

2.4 Other isotherms

In practice it has been found that not all adsorptions behave according to the Langmuir isotherm. Figure 2.9 shows experimental data for adsorption of nitrogen on different solids.

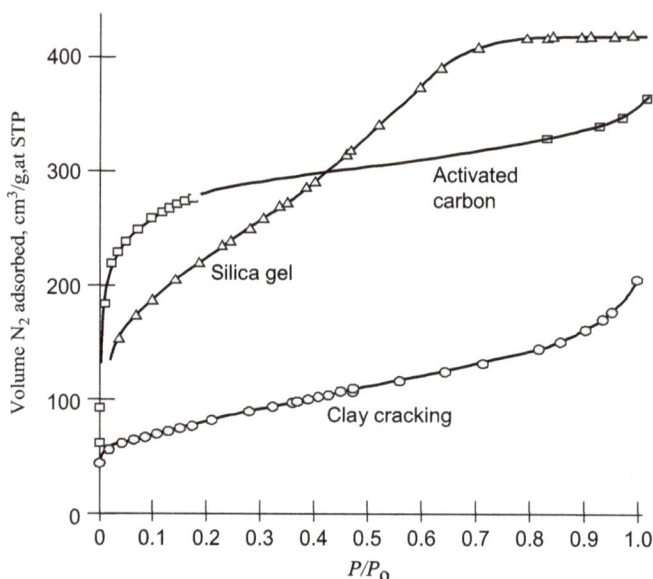

Figure 2.9: Adsorption of nitrogen at −195 °C on porous substances [1].

There are several reasons for the deviation. Perhaps the most important reason is the assumption of energetic equivalence of all sites. It is hard to accept this assumption as the sites themselves are generally not identical. Other assumptions are also difficult to justify. Thus, interactions between adsorbed molecules are possible, as well as multilayer adsorption.

We may then consider the Langmuir isotherm as an *ideal* isotherm, as we do the ideal gas law in comparison to other equations of state.

Several other isotherms were therefore derived; some have a theoretical basis while others are empirical and have been successfully used to fit nonideal systems. Masel [7] and Thomas and Thomas [8] provide comprehensive listings of several isotherms.

Examples of these other isotherms include the **Tempkin** isotherm which is an empirical formula given by the following expression:

$$\theta = c_1 \ln(c_2 P) \tag{2.21}$$

where c_1 and c_2 are constants. Another example is the **Freundlich** isotherm which is given by the following expression:

$$\theta = c_3 P^{1/c_4} \tag{2.22}$$

where c_3 and c_4 are constants. Note that both isotherms predict no finite limiting value for θ. Clearly this is an unrealistic limit.

Example 2.1: Measurements for the adsorption of NO_2 on charcoal were made by McBain and Britton in 1930, in terms of P versus V, where V is the volume of NO_2 adsorbed per g of charcoal, as given below:

P (atm)	0.6	4.0	9.7	19.8	38.0
V (mm^3/g)	0.1525	0.1891	0.1971	0.2064	0.2079

Does the Freundlich isotherm represent the data accurately?

Solution: Let us first write the isotherm in terms of V and P and recast the isotherm into a linear form:

$$\theta = \frac{V}{V_m} = c_3 P^{1/c_4}$$

$$\ln(V) = \ln(V_m c_3) + \frac{1}{c_4} \ln(P)$$

A plot of $\ln(V)$ versus $\ln(P)$ is shown below from which we may conclude that the Freundlich isotherm does not represent the given data accurately.

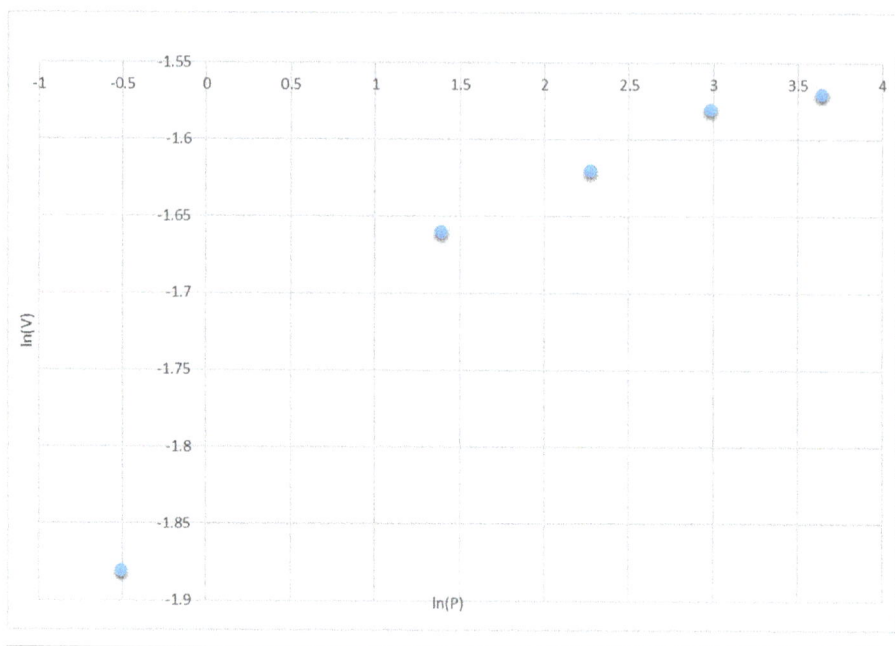

2.4.1 BET isotherm

Brunauer, Emmet, and Teller [9] examined many real adsorption systems and noted that experimentally observed isotherms follow one of the **five** types (shapes) shown in Figure 2.10, as I–V, where plots of the amount adsorbed versus adsorbate partial pressure (normalized by dividing by the vapor pressure P_0 at the temperature in question) are shown.

Later, one more shape (graph VI in Figure 2.10) has been identified. This led to a new *IUPAC classification*[3] of adsorption isotherms. Reference [10] provides a discussion of the background of the new classification.

The limitations of the Langmuir isotherm (what type does it follow?) encouraged Brunauer and colleagues [11][4] to derive an important relation that can predict all five types. This relation is known as the **BET isotherm**. The BET isotherm takes multilayer adsorption into account and its importance is such that it is the basis of *standard* methods for determining the specific surface area of solid catalysts as we shall see later.

---

3 IUPAC is the world authority on chemical nomenclature, terminology, standardized methods for measurement, atomic weights, and other critically evaluated data (see www.iupac.org).
4 This paper is one of the most cited in the physical sciences. For a history of this work, see [13].

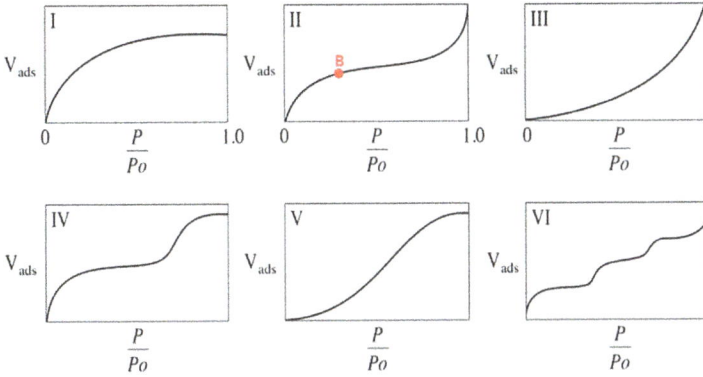

Figure 2.10: The main categories of adsorption isotherms (I–VI) according to the IUPAC classification.

The **general** form of the BET isotherm is given by the following equation:

$$\frac{V}{V_m} = \frac{c(P/P_0)\left[1 - (n+1)(P/P_0)^n + n(P/P_0)^{n+1}\right]}{(1 - P/P)\left[1 + (c-1)(P/P_0) - c(P/P_0)^{n+1}\right]} \tag{2.23}$$

where n is the number of adsorbate layers (do not confuse this n with that n used in deriving the Langmuir isotherm) and c is a temperature-dependent constant. It should be noted that by appropriate choice of c and n, it is possible to generate the five types of isotherms.

For example, when $n = 1$, eq. (2.23) reduces to the Langmuir isotherm. Try this simple exercise and see what happens. On the other hand, when $n \to \infty$, we obtain the **standard** BET isotherm, which is given by the following equation:

$$\frac{V}{V_m} = \frac{c\,P}{(P_0 - P)\,[1 + (c-1)\,(P/P_0)]} \tag{2.24}$$

Before we end this section, let us return to Figure 2.10 and briefly summarize what has been observed about the five types of adsorption isotherms. The summary is given in Table 2.2.

Table 2.2: The main features of the five adsorption isotherm categories.

Type	Remarks
I	Langmuir isotherm; behavior expected in chemisorption
II	Very common; behavior normally observed in physical adsorption. Point **B** corresponds to monolayer capacity
IV	Common
III and V	Relatively rare

2.4.2 Surface area by BET isotherm

While we may use the Langmuir isotherm, as explained earlier, to obtain a rough estimate of the surface area of a heterogeneous catalyst, the standard form of the BET isotherm is the basis of standard ASTM methods[5] for that purpose; namely ASTM D 3663-99 and ASTM D 4567-99. Let us see how we go about calculating the specific area on this basis.

In linear form, eq. (2.24) becomes:

$$\frac{(P/P_0)}{V[1-(P/P_0)]} = \frac{1}{V_m c} + \frac{(c-1)}{V_m c}(P/P_0) \tag{2.25}$$

When a plot of the form shown in Figure 2.11 is constructed, a value of V_m can be obtained from the slope and intercept. Equations (2.19) and (2.20) can then be used to calculate the specific area.

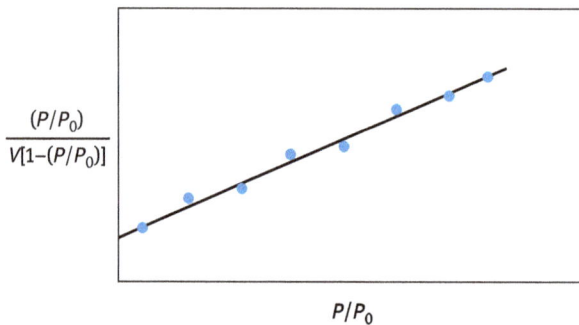

Figure 2.11: A schematic representation of the BET isotherm in linear form.

References

[1] Smith J. M. "Chemical Engineering Kinetics," Third Ed., McGraw-Hill, New York, (1981).
[2] Kenvin J. "Optimizing Catalyst Performance to Support Sustainability Goals," Chem. Eng. Prog., 23–30 (2021).
[3] Fogler H. S. "Elements of Chemical Reaction Engineering," Fifth Ed., Prentice Hall, New Jersey, (2016).
[4] Langmuir I. J. "The Constitution and Fundamental Properties of Solids and Liquids. Part I. Solids," J. Amer. Chem. Soc., 38, 2221 (1916).
[5] Langmuir I. J. "The Adsorption of Gases on Plane Surfaces of Glass, Mica and Platinum," J. Amer. Chem. Soc., 40, 1361 (1918).

---

5 ASTM stands for the American Society for Testing and Materials. It is instructive to visit www.astm. org and look up the two methods.

[6]   Farrautto R. J., and Bartholomew C. H. "Fundamentals of Industrial Catalytic Processes," Blackie Academic & Professional, London, (1997).

[7]   Masel R. I. "Principles of Adsorption & Reaction on Solid Surfaces," Wiley, New York, (1996).

[8]   Thomas J. M., and Thomas W. J. "Principles and Practice of Heterogeneous Catalysis," VCH, Weinheim, (1996).

[9]   Brunauer S., Emmet P. H., and Teller E. "The Adsorption of Gases in Multimolecular Layers," J. Amer. Chem. Soc., 60, 309 (1938).

[10]   Sing K. S. W. "Reporting Physisorption Data for Gas/solid Systems with Special Reference to the Determination of Surface Area and Porosity," Pure Appl. Chem., 54, 2201 (1982).

[11]   Brunauer S., Deming L. S., Deming W. E., and Teller E. "On a Theory of the van der Walls Adsorption of Gases," J. Amer. Chem. Soc., 62, 1723 (1940).

[12]   Butt J.B. "Reaction Kinetics and Reactor Design", Second Ed., Marcel Dekker, New York (2000).

[13]   Davis B. H. "The BET Equation – Nominated for a Nobel Prize but Not Selected," ACS Symp. Series, 1262, 165–206 (2017).

3 Characterization of solid catalysts

Chemical engineers like to stay in their "silos"!
Common misconceptions about chemical engineers, IChemE, UK

In the last part of Chapter 2, we studied how adsorption isotherms could be used for estimating the specific surface area of solid catalysts. There are several other properties that are equally important.

Those properties can be broadly classified as physical, chemical, surface, and mechanical properties. The measurement of catalyst properties or characteristics is called *characterization*.

Catalyst characterization has several objectives. Some of the most important objectives from an engineering point of view include [1]:

(a) To understand the relationship between these properties and the catalyst activity.
(b) To understand what causes loss of catalyst activity with usage (this subject will be considered in more detail in Chapter 6).
(c) To determine the physical, chemical, and structural properties of the catalyst in the catalyst selection (among commercial catalyst alternatives); process reactor design; and process optimization.

There have been excellent developments in characterization *methods* over the last few decades, especially in the areas of chemical characterization and microstructural investigations (alternatively called *surface morphology*).

However, it is beyond the scope of this book to discuss all characterization methods. It is possible, however, to provide a "flavor" of some of the characterization techniques. This is the objective of this chapter. Therefore, we shall look at measurement of few important physical, chemical, structural, and mechanical properties.

3.1 Physical characterization

Physical characterization mainly involves the determination of the *specific surface area*, the *total pore volume*, the *pore-size distribution*, and the *mean pore radius*.

Clearly, the size as well as the number of pores is directly related to the internal surface area. As mentioned in the introductory chapter, a large surface area is generally required in order to maximize the dispersion of the catalytic agent(s) on the support. Likewise, the pore-size distribution is an important property in evaluating how fast reactant molecules can access the active sites.

https://doi.org/10.1515/9783111032511-003

3.1.1 Determination of surface area, total pore volume, and mean pore radius

As explained in Chapter 2, measuring the total surface area of a solid involves the physical adsorption of a gas and determining the number of gas molecules required to cover the surface of the solid with a monolayer of the adsorbate.

If the area occupied by one molecule is known, the surface area of the solid can be calculated from the number of adsorbed gas molecules as measured volumetrically. Adsorption of a gas on a solid is usually characterized by an *isotherm*.

The most common technique for determining the internal surface area of a porous material is the **BET method**[1] based on the adsorption and condensation of N_2 at the temperature of liquid N_2 (i.e., the boiling temperature of N_2). The specific steps involved were given in Section 2.4.2 and therefore will not be repeated here.

At this point, it is worthwhile to mention that, because of the importance of solid catalysts and adsorbents in industry, highly sophisticated tools are available in the scientific instruments market to provide characteristics of both types of solids.

Some of those instruments can be used to collect *and* analyze experimental N_2-adsorption data. Such a commercial unit is shown in Figure 3.1a; typical data collected are shown in Figure 3.1b.

The analysis of the data collected by this instrument includes values of the *specific total pore volume* (also called *void volume*). Typical results are shown in Figure 3.2.

Measuring the pore volume allows us to calculate other properties, namely, the mean pore radius $\bar{r}_p$ and the porosity of the solid particle ε_p, as shown below.

The measured pore volume can be combined with the measured surface area to *estimate* the mean pore size ($\bar{r}_p$). Assuming that the pores are composed of *nonintersecting, uniform cylindrically shaped* channels, the average pore radius is calculated according to

$$\bar{r}_p = \frac{2V_g}{S_g} \tag{3.1}$$

The porosity of the solid particle (also called the *void fraction*) can be calculated according to

$$\varepsilon_p = \frac{V_{void}}{V_{total\ solid}} \tag{3.2}$$

---

1 The BET method has been subject to refinements over the years, such as the alternative use of the BJH method. Specialists should rely on the latest IUPAC guidelines; see [2]. However, for our purposes, the BET method is sufficient.

(a)

(b)

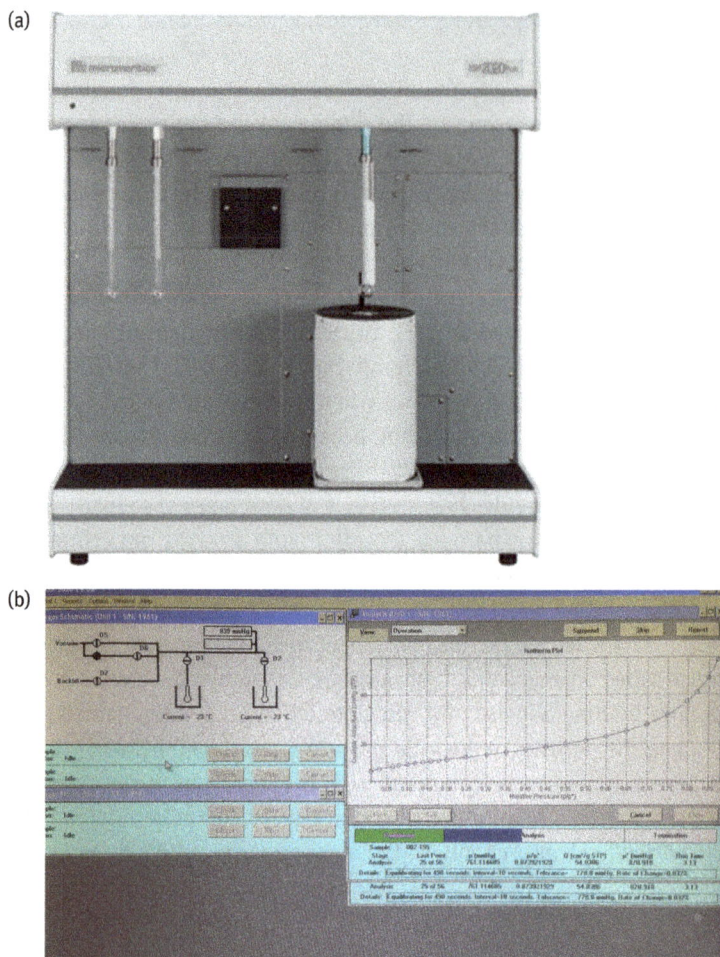

Figure 3.1: (a) Micromeritics® ASAP 2020 unit for surface area determination. (b) A screenshot of experimental data generated by Micromeritics® ASAP 2020 unit for determination of catalyst surface area and mean pore radius (courtesy of Research Institute, KFUPM).

micromeritics®

ASAP 2020 V4.02 (V4.02 H)   Unit 1   Serial #: 1961   Page 1

Sample:
Operator: REY
Submitter: UOP
File: C:\2020\DATA\DATA10\002-770.SMP

Started: 4/27/2021 2:43:01PM
Completed: 4/27/2021 11:18:31PM
Report Time: 4/28/2021 9:29:21AM
Sample Mass: 0.2113 g
Cold Free Space: 48.2947 cm³
Ambient Temperature: 22.00 °C
Automatic Degas: Yes

Analysis Adsorptive: N2
Analysis Bath Temp.: -195.824 °C
Thermal Correction: No
Warm Free Space: 16.4071 cm³ Measured
Equilibration Interval: 10 s
Low Pressure Dose: None

Summary Report

Surface Area

Single point surface area at p/p° = 0.301689453: 130.1257 m²/g

BET Surface Area: 133.8857 m²/g

t-Plot Micropore Area: 0.9635 m²/g

t-Plot External Surface Area: 132.9221 m²/g

BJH Adsorption cumulative surface area of pores
between 17.000 A and 3000.000 A diameter: 154.920 m²/g

BJH Desorption cumulative surface area of pores
between 17.000 A and 3000.000 A diameter: 176.7766 m²/g

Pore Volume

Single point adsorption total pore volume of pores
less than 23519.498 A diameter at p/p° = 0.999185463: 0.610147 cm³/g

t-Plot micropore volume: -0.000331 cm³/g

BJH Adsorption cumulative volume of pores
between 17.000 A and 3000.000 A diameter: 0.609756 cm³/g

BJH Desorption cumulative volume of pores
between 17.000 A and 3000.000 A diameter: 0.611226 cm³/g

Pore Size

Adsorption average pore width (4V/A by BET): 182.2889 A

BJH Adsorption average pore diameter (4V/A): 157.437 A

BJH Desorption average pore diameter (4V/A): 138.305 A

Pass/Fail

S A:Multi-point BET Failed: Value < lowest allowed 219.0000 by 85.1143
S A:Multi-point BET Failed: Value < lowest allowed 219.0000 by 85.1143
S A:Single-point BET Failed: Value < lowest allowed 204.0000 by 73.8743
S A:Single-point BET Failed: Value < lowest allowed 204.0000 by 73.8743
P V:Adsorption Total - 0.99500000 Relative Pressure Passed
P S:Avg. pore diameter Failed: Value > highest allowed 131.000 by 51.289
P S:Avg. pore diameter Failed: Value > highest allowed 131.000 by 51.289

Figure 3.2: Typical analysis of experimental data generated by Micromeritics® ASAP 2020 unit (courtesy of Dr. Z. Qureshi, Research Institute, KFUPM).

Table 3.1 gives typical values of pore volume and pore radii for some common solids.

Table 3.1: The surface area, pore volume, and mean pore radii for some porous solids [3].

Solid	S_g (m²/g)	V_g (cm³/g)	$\bar{r}_p$ (nm)
Activated carbons	500–1,500	0.6–0.8	1–2
Silica gels	200–600	0.4	1.5–10
Activated clays	150–225	0.4–0.52	10

Table 3.1 (continued)

Solid	S_g (m²/g)	V_g (cm³/g)	$\bar{r}_p$ (nm)
Activated alumina	175	0.39	4.5
Kieselguhr[a]	4.2	1.1	1,100
NH₃ catalysts (Fe₂O₃)	–	0.12	20–100

[a]Kieselguhr, also known as *diatomaceous earth* and *Celite*, is a naturally occurring, chalk-like sedimentary rock that is easily crumbled into a fine powder (see Figure 3.3). This powder has an abrasive feel and is very light due to its high porosity. Its typical chemical composition is 86% silicon, 5% sodium, 3% magnesium, and 2% iron.

Figure 3.3: A picture of raw Kieselguhr powder.

Example 3.1: The following data were collected by Holland and Murdoch [4], at −195.2 °C, for a Cr/Al₂O₃ catalyst by using a N₂-adsorption apparatus. The volume of N₂ adsorbed is reported at 0 °C and 1 atm. Estimate the specific surface area.

P (cm Hg)	1.145	18.130	32.870	41.640	46.900
V (cm³/g)	11.38	18.17	23.07	26.85	29.88

Solution: The linear form of the BET isotherm is given by eq. (2.25). A plot of the data is shown below, where $P_0 = 81.11$ mm Hg was used. Using the plot, we find that:

Slope = 0.072
Intercept ≈ 0
Therefore, $V_m = 13.33$ cm³/g.
Using eq. (2.19), we find that $S_g = 58.06$ m²/g.

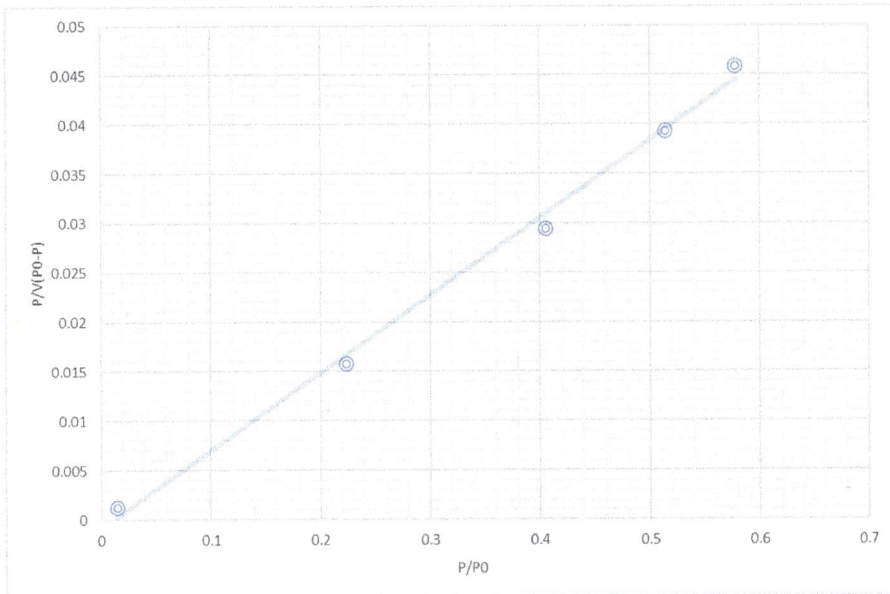

3.1.2 Pore-size distribution

When the catalyst structure is examined at the microscopic scale, one normally finds that the pores are not uniform in size, shape, or length, except in the special case of zeolites.[2]

Figure 3.4 shows typical void spaces in a specific functional material[3] particle. In this figure, a variety of void spaces, some much larger than others, can be seen. Therefore, we expect a *distribution* of pore sizes in typical catalyst particles. On this basis, pores have been classified by IUPAC into three categories:

Macropores: $r_p > 25$ nm;   or $d_p > 50$ nm
Mesopores: $1 < r_p < 25$ nm;   or $2 < d_p < 50$ nm
Micropores: $r_p < 1$ nm;   or $d_p < 2$ nm

The distribution of pore sizes can be obtained by two standard ASTM methods. One technique involves the physical adsorption–desorption isotherm for N_2. The second technique is based on the *mercury-intrusion porosimetry.*

---

2 Zeolites are the aluminosilicate members of a family of microporous solids known as "molecular sieves." A particular property of these materials is the very regular pore structure.
3 Functional materials include materials such as solid catalysts, semiconductors over polymers; and nanoparticles.

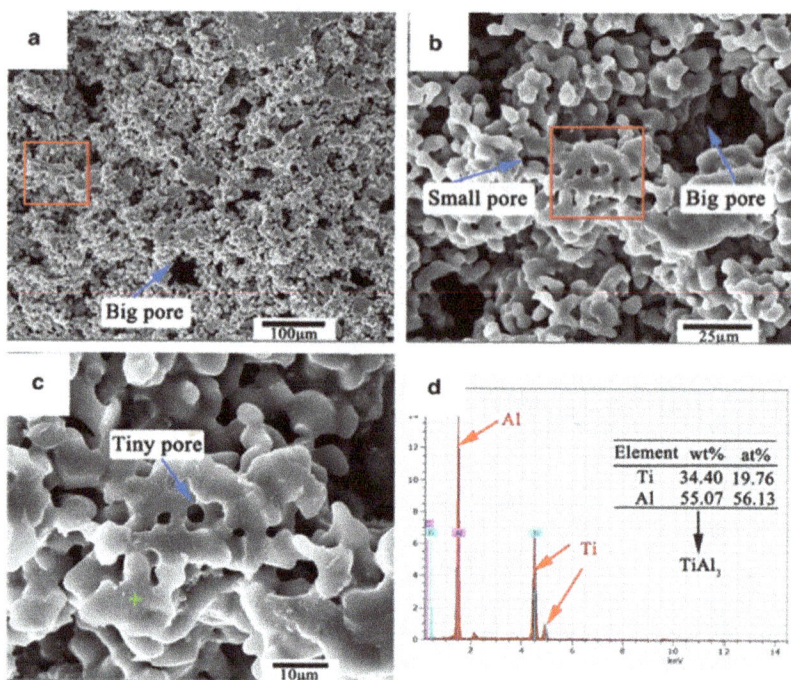

Figure 3.4: Pore morphology of porous TiAl₃ intermetallics: (a) low-magnification SEM image; (b) high magnification of small pores; (c) tiny pores; (d) EDS/SEM analysis for the selected point [5].

The first method is applicable to pores smaller than about 60 nm. The second method is applicable to pores with diameters larger than 3.5 nm [6]. Here we discuss only the second method.

3.1.2.1 Mercury-porosimetry method

As mentioned in the preceding paragraph, this method is applicable to pore diameters larger than 3.5 nm. According to the above classification of pore sizes, the method allows both *macro*pores and *meso*pores to be determined.

The basic idea of this technique is based on the fact that mercury does not "wet" most surfaces. Therefore, Hg will not penetrate small pores unless pressure is applied. The pore-size distribution is determined by measuring the volume of Hg that enters pores under pressure. Figure 3.5 shows typical data.

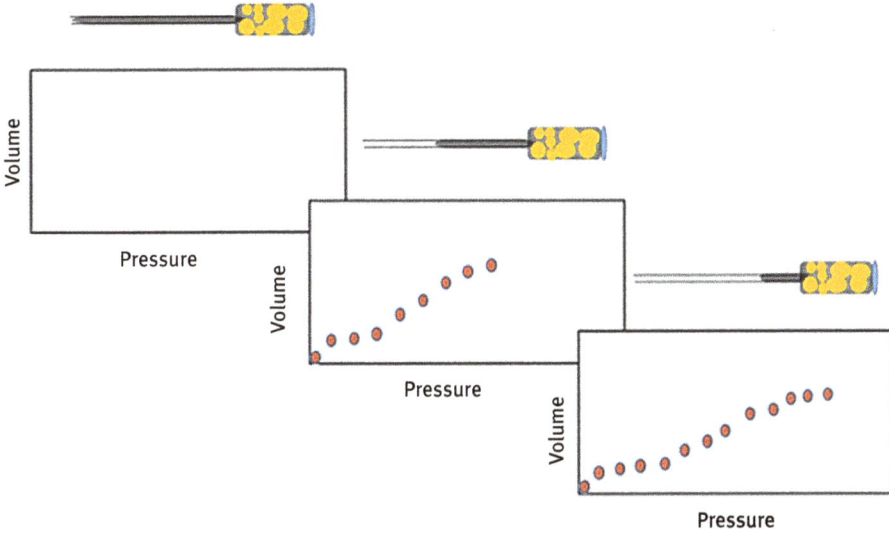

Figure 3.5: A schematic of the progressive operation of the glass sample cell in mercury-intrusion porosimetry.

The penetration of mercury into the pores of a material is a function of the applied pressure according to the *Washburn equation* [7]:

$$P = \frac{2\pi \; \sigma \cos(\alpha)}{r_\mathrm{p}} \qquad (3.3)$$

where
P = measured pressure
α = contact angle between mercury and the solid (130°)
σ = surface tension of mercury
r_p = radius of cylindrical pore

Figure 3.6a shows the Hg-penetration curve for a specific solid. The collected data can subsequently be used to generate a pore-volume distribution curve, as shown in Figure 3.6b, from which the mean pore radius, $\bar{r}_\mathrm{p}$, can be determined as indicated in the same figure.

(a)

(b)

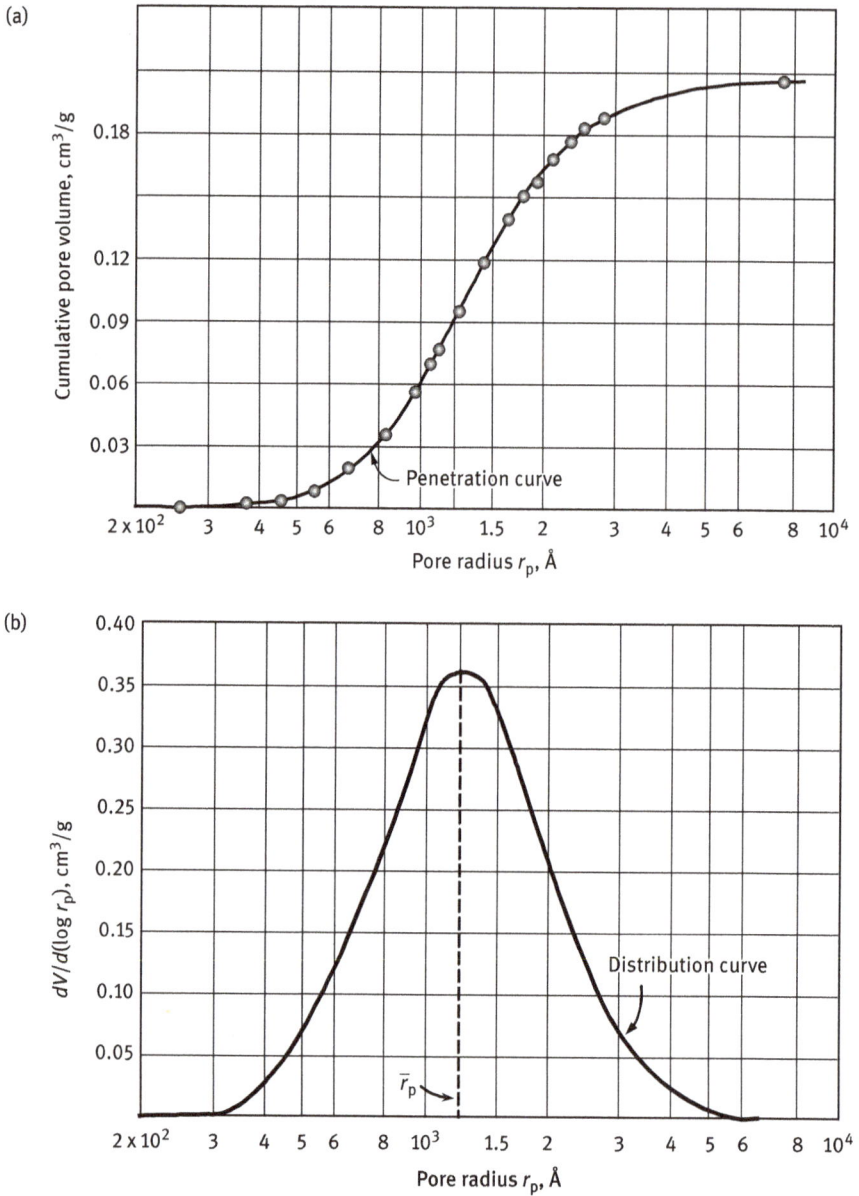

Figure 3.6: (a) Pore-volume penetration data for mercury in a UO_2 catalyst pellet [8].
(b) The pore-volume distribution curve, based on the data in (a) [8]. The *mean pore radius* is
indicated by the dashed line.

3.2 Chemical and structural characterization

Many sophisticated methods have been developed and applied over the last few decades to obtain chemical and microstructural properties of solids.

Most available methods for the characterization of solid catalysts have been cataloged and detailed in Bartholomew and Farrauto [1], Thomas and Thomas [9], Ertl et al. [10], and recently by Ross [11]. A summary of most of the chemical and structural characterization methods is presented in Table 3.2.

Table 3.2: Chemical and structural characterization methods (adapted from [1]).

Characteristic	Method[a]
Bulk chemical state	TG, TPR, MAS, FTIR, EPR, NS, NMR
Chemical state of the surface	XPS, TPSR, EXAFS, MAS, FTIR, NMR
Chemical state of the additives	XPS, FTIR, NMR, EELS, HREELS, TPD, TPSR, MBS
Bulk composition	Elemental analysis, (AAS and ICP), XRF/XRD, SEM/EDX, TEM, MAS, TG, FTIR, NMR
Surface composition	AES, XPS, SIMS, EXAFS, ISS
Surface acidity	Adsorption and TPD of bases, IR, NMR
Surface reactivity, active site concentration	Chemisorption, TPD, TPSR, transient kinetic methods, calorimetry
Bulk structure morphology	XRD, SEM, TEM, STEM, MAS
Surface structure morphology	LEED, EXAFS, TEM, STEM, STM, FEM
Dispersion	Chemisorption, TEM, XRD

[a]The reader is referred to reference [1] for the meaning of these abbreviations (acronyms). Alternatively, they can be looked up at http://en.wikipedia.org/wiki/List_of_surface_analysis_methods.

Chemical properties of catalytic materials include composition, active site concentration, and chemical state. *Structural* properties include the morphology of the solid surface and the texture of the bulk solid.

In general, chemical and structural properties are primarily of interest to the developers and manufacturers of catalysts; these properties are *seldom routinely* determined during catalyst usage unless problems arise in the plant operation: e.g., the catalyst loses its activity prematurely, or the catalyst particles crumble when loaded in the reactor (we are talking about *tons* of the catalyst particles!).

As shown in Table 3.2, most of these properties are determined by *spectroscopic* methods. To give a "flavor" of what is involved, this section focuses on two commonly used methods involving *electron microscopy*.

3.2.1 Electron microscopy

Electron microscopy is a powerful tool for the analysis of chemical elements, surface texture and morphology, and crystallite size of catalysts.

Two types of electron microscopy are widely used: *scanning* electron microscopy **(SEM)** and *transmission* electron microscopy **(TEM)**. Note that SEM and TEM are similar to optical microscopy, where the optical lenses are replaced with electromagnetic ones. Figures 3.7a and b show the principles of operation of SEM and TEM.

SEM involves passing a narrow electron beam through condenser lenses to produce parallel rays over the solid sample surface and detecting the yield of either secondary or backscattered electrons (see Figure 3.7b) as a function of the position of the primary beam [13]. Backscattered electrons carry information that allows surface analysis at a resolution of 10 nm [1].

In TEM, a primary electron beam of high intensity passes also through condenser lenses that impinge on the solid sample. As attenuation of the beam depends on the density and composition of the sample, transmitted electrons form a two-dimensional image of the sample mass, which is subsequently enlarged by the electron optics [13]. TEM allows surface analysis up to a resolution of 0.2–0.3 nm [1].

Although few SEM images were presented in Chapter 2, let us look at some specific micrographs related to the spherical resin catalyst shown in Figure 1.1. As in the case of most industrial catalysts, the resin catalyst used in the production of MTBE is produced and sold by different suppliers.

Figure 3.8 shows SEM/EDX micrographs for three **spent** samples of the same catalyst obtained from different suppliers. Coupled with SEM, energy dispersive X-ray spectroscopy (SEM/EDX) is one of the most widely used surface analytical techniques. This combination allows us to determine catalyst particle size and surface morphology, in addition to elemental composition.

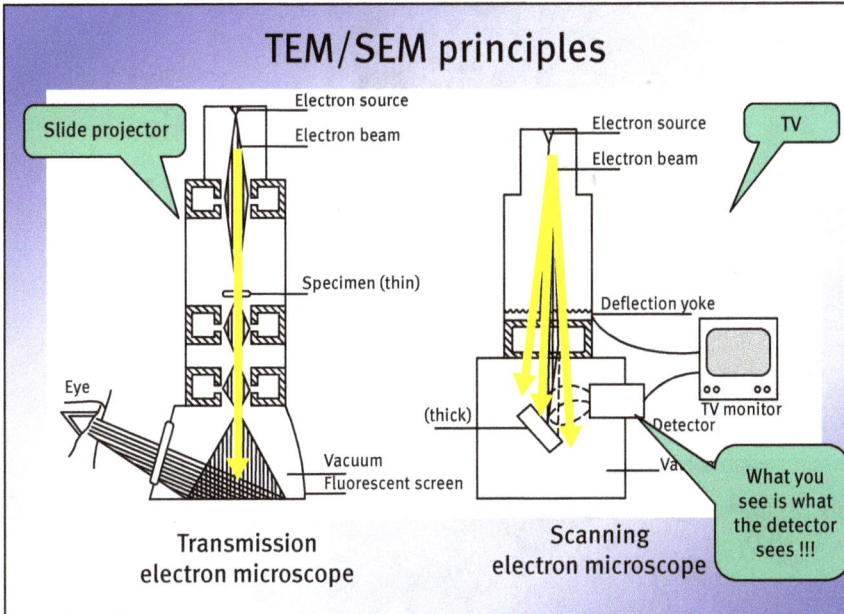

(a)

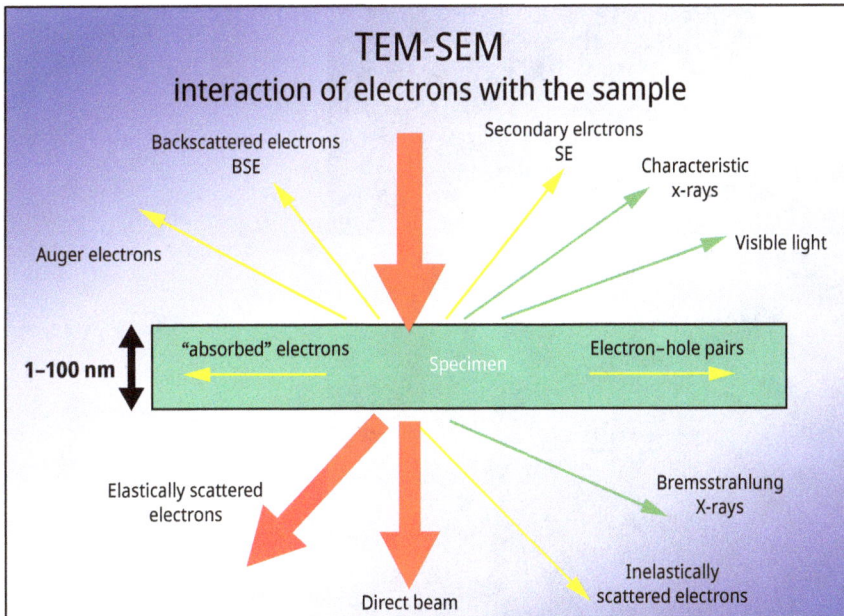

(b)

Figure 3.7: (a) A schematic representation of electron microscopes in the transition (TEM) and scanning modes (SEM) (adapted from [12]). (b) A schematic representation of interactions of primary electron beam (incident) with the solid sample (specimen) in EM, leading to several detectable signals (adapted from [12]).

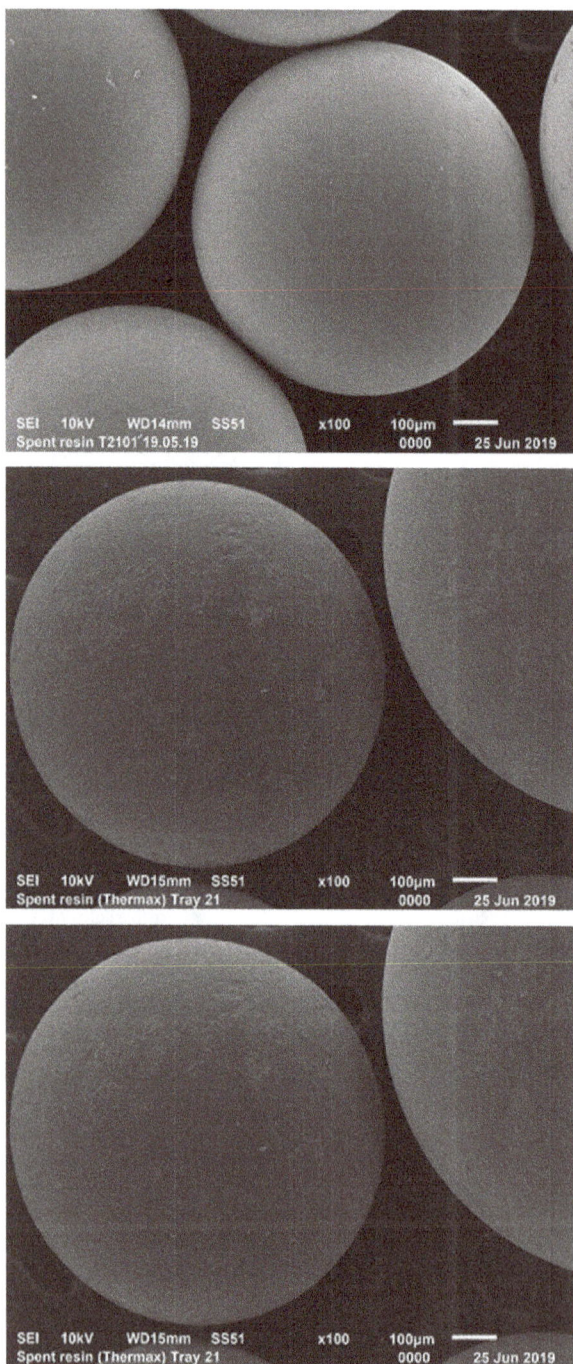

SEI 10kV WD14mm SS51   x100   100µm ▬▬▬
Spent resin T2101 19.05.19   0000   25 Jun 2019

SEI 10kV WD15mm SS51   x100   100µm ▬▬▬
Spent resin (Thermax) Tray 21   0000   25 Jun 2019

SEI 10kV WD15mm SS51   x100   100µm ▬▬▬
Spent resin (Thermax) Tray 21   0000   25 Jun 2019

Figure 3.8: SEM micrographs of three spent resin catalysts (courtesy of Dr. Cristovao de Lemos, Sipchem T&I Center).

3.3 Mechanical characterization

After production, shipping, and loading into industrial reactors (we are still talking about many tons!), solid catalysts could suffer stresses that would reduce their mechanical strength. Why does this happen?

Such stresses generally result from severe operating temperatures and pressures, as well as particle crushing, dusting, and abrasion. These stresses play a role in determining the *lifetime of the catalyst* and economics of the process.

Solid catalysts used in *packed-bed reactors* are subject to the static head of the bed height and should thus be sufficiently strong to resist crushing. The **crushing strength** of catalyst particles (in spherical, tablet, or extrudate form) is determined according to the ASTM D-4,179 method, to which the reader is referred for details of the procedure.

The measurements are simple and consist of a single particle being placed between parallel plates of a device capable of exerting full stress, after which the force required to crush the material is noted. The particles can be placed within the plates to measure axial or radial crush strength [1]. A commercially available machine is shown in Figure 3.9. Figure 3.10 shows typical test results for the crushing strength.

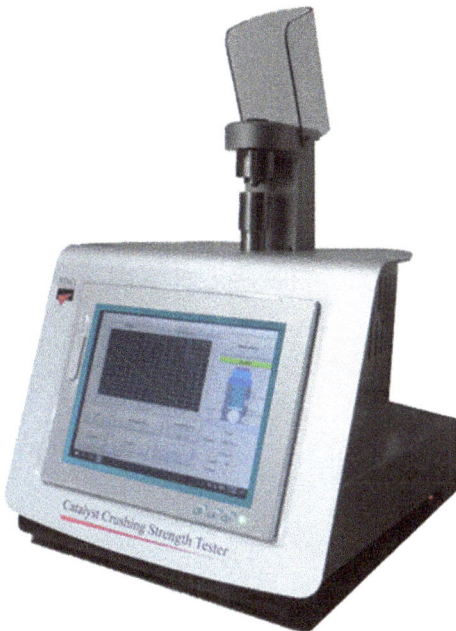

Figure 3.9: Crushing strength testing machine (www.vinci-technologies.com).

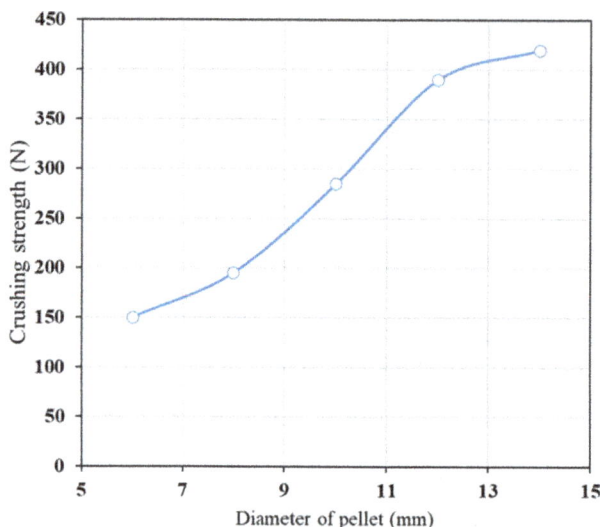

Figure 3.10: Mechanical strength tests for a spherical Ni/Al$_2$O$_3$ catalyst particle.

As we shall see in Chapter 7, there are reactors in which the catalyst particles are not stationary, such as the two-phase *fluidized-bed reactor* and the three-phase *slurry reactor*.

In those reactors, the catalyst particles are subject to abrasion resulting from particle-to-particle collisions as well as collisions with the reactor walls and internals. **Attrition** and **abrasion resistance** thus are important catalyst properties. Here again, a standard ASTM test (D-4,058) is available for measuring these properties.

The test involves placing the catalyst (tablets, extrudates, spheres, or irregularly shaped catalysts) in a drum that rotates for a given amount of time and from which the produced **"fines"** are measured. A commercially available machine is shown in Figure 3.11. Typical abrasion test results are shown in Figure 3.12.

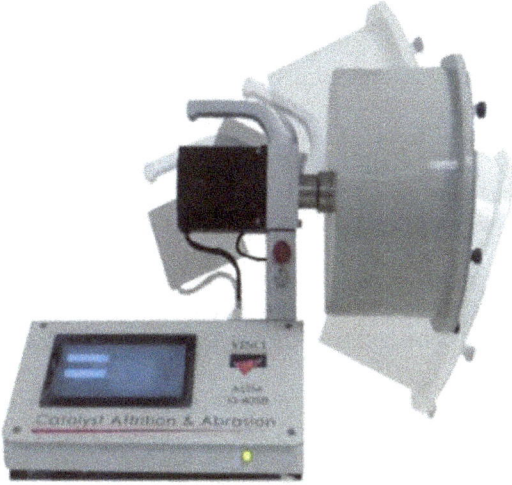

Figure 3.11: Abrasion testing machine (www.vinci-technologies.com).

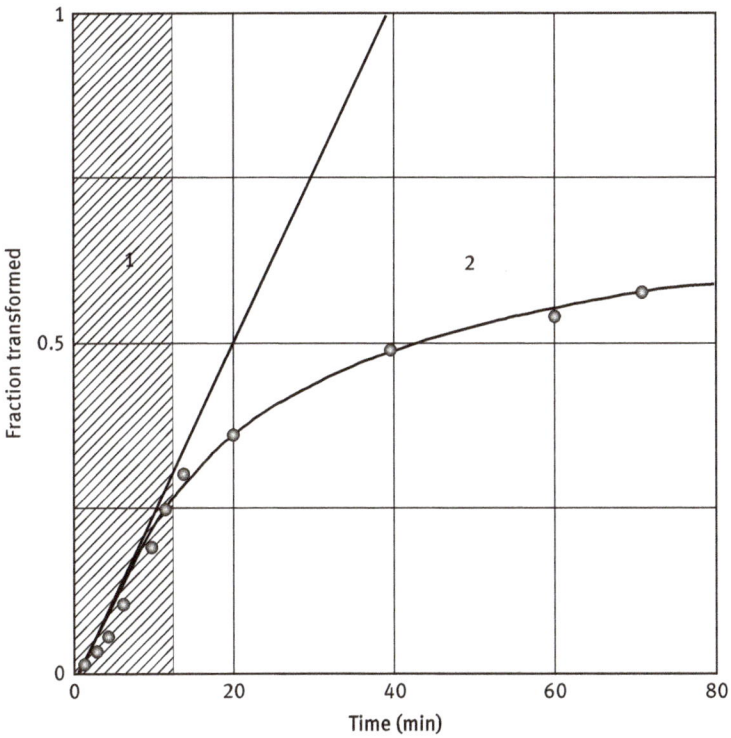

Figure 3.12: Abrasion tests for cylindrical grains: results in stage 1 are faster than those in 2; cited in [6].

References

[1]   Bartholomew C. H., and Farrauto R. J. "Fundamentals of Industrial Catalytic Processes," Second Ed., Wiley, New York, (2006).

[2]   Thommes M., et al. "Physisorption of Gases, with Special Reference to the Evaluation of Surface Area and Pore Size Distribution (IUPAC Technical Report)," Pure Appl. Chem., 87, 1051 (2015).

[3]   Lee H. H. "Heterogeneous Reactor Design," Butterworth, Boston, (1985).

[4]   Holland C. D., and Murdoch P. G. "Determination of the Chemical Properties of an Oxide Catalyst in a Closed System," AIChE J., 3, 386 (1975).

[5]   Jiao X., et al. "Microstructure Evolution and Pore Formation Mechanism of Porous $TiAl_3$ Intermetallics via Reactive Sintering," Acta. Met. Sinica. (English Letters), 31, 440 (2017).

[6]   Le Page J. F. "Applied Heterogeneous Catalysis – Design, Manufacture, and Use of Catalysts," Ēdition Technip, Paris, (1987).

[7]   Hagen J. "Industrial Catalysis – A Practical Approach," Wiley-VCH, Weinheim, (1999).

[8]   Smith J. M. "Chemical Engineering Kinetics," Third Ed., McGraw-Hill, New York, (1981).

[9]   Thomas J. M., and Thomas W. J. "Principles and Practice of Heterogeneous Catalysis," VCH, Weinheim, (1997).

[10]   Ertl G., Knozinger H., and Weitkamp J. "Handbook of Heterogeneous Catalysis," Vol. 2, Wiley-VCH, Weinheim, (1997).

[11]   Ross J. R. H. "Contemporary Catalysis: Fundamentals and Current Applications," Elsevier, Amsterdam, (2019).

[12]   Courtesy of CIME (Interdisciplinary Centre for Electron Microscopy), at the Ēcole Polytechnique Fédérale De Lausanne, Switzerland (http://cime.epfl.ch)

[13]   Chorkendorff I., and Niemantsverdriet J. W. "Concepts of Modern Catalysis and Kinetics," Wiley-VCH, Weinheim, (2003).

4 Kinetics of catalytic reactions

All generalizations are wrong . . . including this one!
Ghazi Al-Gosaibi

As was explained in Chapter 1, the *global catalytic sequence* consists of several physical and chemical steps. These include the transport of species from the bulk fluid phase to the surface of the catalyst (and transport of products from the surface to the bulk), the transport of species through the porous structure (and the similar transport of products), and the *catalytic reaction cycle*. It has also been mentioned that the reaction cycle itself consists of adsorption, surface reaction, and desorption steps.

One or more of those steps may be the rate-limiting step (RLS). In what follows, our main interest is to determine the *intrinsic* kinetics of the surface reaction. Therefore, the entire catalytic surface inside the pellet is assumed to be exposed to a reactant of uniform concentration and temperature, which means that the effects of mass (and heat) transfer *to* or *within* the porous catalyst are considered unimportant. We will consider those effects in the next chapter.

4.1 Classification of kinetic models

In general, models for the kinetics of surface reactions fall into one of four categories: power law, Langmuir–Hinshelwood (L-H), Eley–Rideal (E-R), and Mars–van Krevelen (M-vK) models.

Power-law kinetics is familiar to us because it is commonly used to represent the kinetics of homogeneous reactions. It is thus the simplest approach in which the reaction rate is expressed in terms of a power law of the type:

$$(-r_A) = k \ C_A^{\alpha} \ C_B^{\beta} \tag{4.1}$$

where k, α, and β are empirically determined coefficient and constants. An example of such kinetics is given in [1].

Yet, over the years, the L-H and E-R models have been widely used to provide mechanistic descriptions of catalytic reactions.[1] We consider these models in more detail in the next sections.

The fourth model (M-vK) has been applied mainly to describe the kinetics of selective oxidations of hydrocarbons [2]. Therefore, we shall not discuss the M-vk model any further.

---

1 A mechanistic description is a useful framework for analysis, but it does not necessarily reflect the true mechanism of the reaction.

https://doi.org/10.1515/9783111032511-004

4.2 Langmuir–Hinshelwood (L-H) model

The L-H approach is based on the suggestion advanced by Langmuir in relation to adsorption equilibria.

Originally, the L-H models were expressed in terms of surface coverage, but later Hougen and Watson [3] reorganized the L-H analysis in terms of active site concentrations. (The L-H model is thus called the LHHW, particularly in the chemical engineering literature.) Conceptually, there is no difference between the L-H and H-W formulations except that the latter is more convenient in deriving kinetic expressions.

The L-H approach will be used next to derive a few illustrative kinetic models, although many more can be derived (see Butt [4] for other reactions).

4.2.1 Irreversible, unimolecular, surface reactions

Consider the simple catalytic reaction

$$A_{(g)} \xrightarrow{\text{catalyst}} B_{(g)}$$

If this reaction is assumed to involve two adsorbed molecules of the reactant and product on the **active site X**, then a possible mechanism is

$$A + X \Leftrightarrow AX$$

$$AX \rightarrow BX$$

$$BX \Leftrightarrow B + X$$

Assuming that the second step is the RLS, and the reaction rate is proportional to the fraction of surface covered, then according to the expressions derived in Chapter 2 for the fraction of surface covered by species A and B (based on the Langmuir isotherm for competitive adsorption), we can express the reaction rate as:

$$(-r_A) = k\theta_A = \frac{k \, K_A \, C_A}{1 + K_A \, C_A + K_B \, C_B} \tag{4.2}$$

Two special cases can be recognized: the first is when the product is weakly adsorbed and the second is when the product is strongly adsorbed.

Special case: Weakly adsorbed product ($K_B C_B \ll 1$)

$$(-r_A) = k \, \theta_A = k \frac{K_A C_A}{1 + K_A C_A} \tag{4.3}$$

The rate behavior is depicted schematically in Figure 4.1, where we can see that the rate is invariant at large concentrations of the reactant. Examples of reactions following

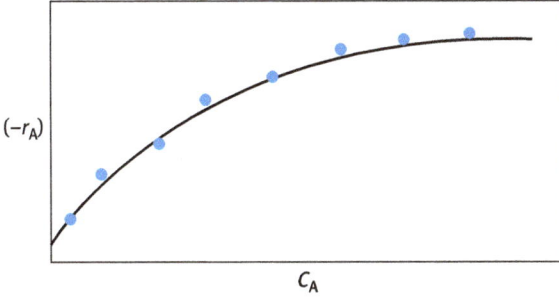

Figure 4.1: The behavior of eq. (4.3).

this behavior include the decomposition of ammonia over tungsten, and the decomposition of phosphine over glass [5].

Special case: Strongly adsorbed product ($K_B C_B \gg 1$)

$$(-r_A) = \frac{k\ K_A\ C_A}{K_B\ C_B} \tag{4.4}$$

In this case, the reaction is inhibited by the strongly adsorbed B. Note that the observed rate is first-order with respect to the reactant and inverse first-order with respect to the product. This behavior has been observed in, for example, the decomposition of ammonia over platinum [5].

4.2.2 Irreversible, bimolecular, surface reactions

Consider the bimolecular catalytic reaction

$$A + B \rightarrow R + Q$$

The L-H mechanism assumes that the reaction occurs between molecules of species A and B when both are adsorbed on the surface. Assuming that the reaction rate is proportional to θ_A and θ_B, the rate can be expressed as

$$(-r_A) = k\ \theta_A\ \theta_B = k \cdot \frac{K_A\ C_A}{1 + K_A\ C_A + K_B\ C_B} \cdot \frac{K_B C_B}{1 + K_A C_A + K_B C_B} \tag{4.5}$$

based on surface coverage expressions for competitive adsorption, or

$$(-r_A) = \frac{k\ K_A\ K_B\ C_A\ C_B}{[1 + K_A\ C_A + K_B C_B]^2} \tag{4.6}$$

The behavior of the rate equation is schematically shown in Figure 4.2 for the case of constant C_B. Note that the rate first increases, passes through a maximum, and finally

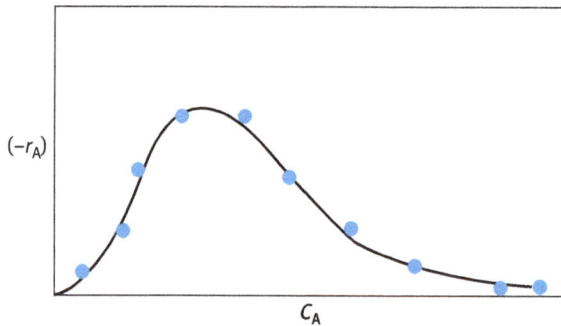

Figure 4.2: The behavior of eq. (4.6) at constant C_B.

decreases with increasing values of C_A. The reason for the decrease in the reaction rate at high C_A values is that one reactant displaces the other as its concentration is increased.

Special case: Sparsely covered surface

If C_A and C_B are both sufficiently low such that $K_A C_A \ll 1$ and $K_B C_B \ll 1$, eq. (4.6) reduces to:

$$(-r_A) = k\, K_A\, K_B\, C_A\, C_B \tag{4.7}$$

The reaction is first-order with respect to A and B, that is, second-order overall.

Special case: One reactant weakly adsorbed

If reactant A is weakly adsorbed such that $K_A C_A \ll 1$, the rate equation reduces to:

$$(-r_A) = \frac{k\, K_A\, K_B\, C_A\, C_B}{(1 + K_B\, C_B)^2} \tag{4.8}$$

Here, the rate still passes through a maximum as C_B *increases*, but as long as the condition $K_A C_A \ll (1 + K_B C_B)$ remains satisfied, the rate is proportional to C_A. This behavior has been observed in, for example, the reaction of carbon dioxide and hydrogen over platinum [5].

Special case: One reactant strongly adsorbed

If $K_B C_B \gg (1 + K_A C_A)$, the rate equation becomes:

$$(-r_A) = \frac{k\, K_A K_B\, C_A C_B}{(1 + K_A C_A + K_B C_B)^2} \rightarrow \frac{k\, K_A\, K_B\, C_A\, C_B}{(K_B\, C_B)^2}\ \text{or}$$

$$(-r_A) = \frac{k\, K_A\, C_A}{K_B C_B} \tag{4.9}$$

The rate is now inversely proportional to C_B, that is, the order with respect to species B is -1. This peculiar behavior has been observed in the reactions of carbon monoxide and hydrogen over platinum-based and quartz-based solid catalysts, with respect to the concentration of carbon monoxide.

Do you remember such negative orders when you studied *homogeneous* reactions?

4.2.3 Inhibition in bimolecular surface reactions

Consider the irreversible reaction

$$A + B \xrightarrow{\text{catalyst}} \text{Products}$$

If a substance other than A and B is present in the system (a situation especially common in industrial feeds), the reaction could be inhibited because of some of the active sites being occupied by that substance.

Let us denote this substance by "I"; this substance can simply be an *inert* species or a so-called *poison* (poisoning as a catalyst *deactivation* mechanism is discussed in more detail in Chapter 6).

If K_I represents the adsorption constant for species I, then, based on analyses similar to those in Chapter 2, the surface coverage can be represented by the following equations:

$$\theta_A = \frac{K_A\,C_A}{1 + K_A C_A + K_B C_B + K_I C_I} \tag{4.10a}$$

$$\theta_B = \frac{K_B C_B}{1 + K_A C_A + K_B C_B + K_I C_I} \tag{4.10b}$$

The reaction rate expression based on the above equations becomes:

$$(-r_A) = \frac{k\,K_A\,K_B\,C_A\,C_B}{\left[1 + K_A C_A + K_B C_B + K_I C_I\right]^2} \tag{4.11}$$

If species I is strongly adsorbed compared to A and B, then eq. (4.11) reduces to the simple form:

$$(-r_A) = \frac{k\,K_A\,K_B\,C_A\,C_B}{K_I^2\,C_I^2} \tag{4.12}$$

Here, it is clear that the presence of this species decreases or inhibits the rate of the catalytic reaction.

4.2.4 Reversible, unimolecular, reactions

Let us consider the general case in which we have a single species reacting reversibly in the presence of a solid catalyst according to A $\Leftrightarrow$ B. A possible mechanism in the present case is given by:

$$A + X \Leftrightarrow AX$$

$$AX \Leftrightarrow BX$$

$$BX \Leftrightarrow B + X$$

Assuming the surface reaction is the RLS, the net rate of reaction is:

$$(-r_A) = k_1\theta_A - k_2\theta_B \tag{4.13}$$

where k_1 and k_2 are reaction rate coefficients for the forward and reverse reactions, respectively. In terms of surface concentrations, eq. (4.13) is equivalent to:

$$(-r_A) = k_1 C_{AX} - k_2 C_{BX} \tag{4.14}$$

Therefore, in terms of bulk phase concentrations, we obtain:

$$(-r_A) = \frac{k_1 \, K_A \, C_A}{(1 + K_A C_A + K_B C_B)} - \frac{k_2 \, K_B C_B}{(1 + K_A C_A + K_B C_B)}$$

or

$$(-r_A) = \frac{k_1 \, K_A C_A - k_2 \, K_B C_B}{(1 + K_A C_A + K_B C_B)} \tag{4.15}$$

4.2.5 Reversible, bimolecular, reactions

We now consider the general case in which two species react reversibly in the presence of a solid catalyst according to:

$$A + B \Leftrightarrow R + Q$$

In a way similar to that described in the previous section, a possible mechanism can be written as:

$$A + X \Leftrightarrow AX$$

$$B + X \Leftrightarrow BX$$

$$AX + BX \Leftrightarrow RX + QX$$

$$RX \Leftrightarrow R + X$$

$$QX \Leftrightarrow Q + X$$

Assuming surface reaction control like before, the net reaction rate is:

$$(-r_A) = k_1 \, C_{AX} \, C_{BX} - k_2 \, C_{QX} \, C_{RX} \tag{4.16}$$

For competitive adsorption of A, B, Q, and R on the surface, the Langmuir isotherm gives us the surface coverage in terms of partial pressures:

$$\theta_j = \frac{K_j P_j}{1 + K_A P_A + K_B P_B + K_Q P_Q + K_R P_R} = \frac{K_j P_j}{1 + \sum K_j P_j} \tag{4.17}$$

where j = A, B, Q, or R. Hence the rate of reaction in terms of partial pressures is:

$$(-r_A) = \frac{k_1 \, K_A K_B \, P_A P_B - k_2 \, K_Q K_R \, P_Q P_R}{\left(1 + K_A P_A + K_B P_B + K_Q P_Q + K_R P_R\right)^2} \tag{4.18}$$

4.2.6 Irreversible, bimolecular, reactions: different sites

It has been implicitly assumed in all the previous derivations that the active sites are all identical and that they can be represented by the symbol X.

However, as the solid is not necessarily uniform either in terms of catalyst structure or composition, or in terms of the adsorbed species, it is possible to imagine situations in which *different* catalyst sites are involved in the surface reaction. For example, in the case of metal oxide catalysts, one might argue that certain species could be adsorbed by interaction with metal atoms at the surface, while other species could interact with surface O_2 atoms.

Let us consider by way of illustration the irreversible reaction:

$$A_{(type\,1\,site)} + B_{(type\,2\,site)} \rightarrow Products$$

A possible mechanism for this reaction can be written as follows:

$$A + X_1 \Leftrightarrow AX_1$$

$$B + X_2 \Leftrightarrow BX_2$$

$$AX_1 + BX_2 \rightarrow Products$$

Following the same procedure used in the previous cases, the rate is proportional to the surface coverage on the different sites, in which case we can write:

$$(-r_A) = k \left(\frac{K_{A1} P_A}{1 + K_{A1} P_A} \right) \left(\frac{K_{B2} P_B}{1 + K_{B2} P_B} \right) \tag{4.19}$$

An example of a catalytic reaction with kinetics following eq. (4.19) is the reaction of hydrogen and carbon dioxide over tungsten [5].

At this point, we may well ask ourselves about the difference in the behavior of eq. (4.19) vs. eq. (4.6). Does eq. (4.19) display a maximum?

4.3 Eley–Rideal (E-R) model

To illustrate this mechanistic model let us consider the irreversible bimolecular reaction:

$$A + B \rightarrow R + Q$$

In the E-R model, it is suggested that the catalytic reaction occurs between *molecules of species A in the gas phase* and adsorbed molecules of species B, so that only one reactant has to be adsorbed. It should be noted that this mechanism does not require that A is not adsorbed; it simply postulates that adsorbed A molecules do *not* react. The situation is shown schematically in Figure 4.3.

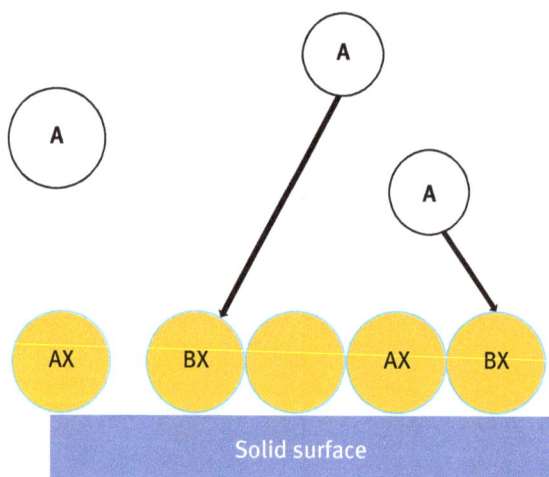

Figure 4.3: A schematic of the E-R model for catalytic reactions.

The mechanism can be written as follows:

$$B_{(g)} + X_{(s)} \Leftrightarrow BX_{(s)}$$

$$A_{(g)} + BX_{(s)} \Leftrightarrow ABX_{(s)}$$

$$ABX_{(s)} \rightarrow R_{(g)} + Q_{(g)} + X_{(s)}$$

Assuming that the reaction occurs between adsorbed molecules of B and molecules of A in the gas phase, the reaction rate is proportional to the fraction of surface covered by B and the concentration of A in the bulk phase, or

$$(-r_A) = k\theta_B C_A = \frac{k\,K_B\,C_A\,C_B}{(1+K_A C_A + K_B C_B)} \tag{4.20}$$

The behavior of eq. (4.20) is shown schematically in Figure 4.4.

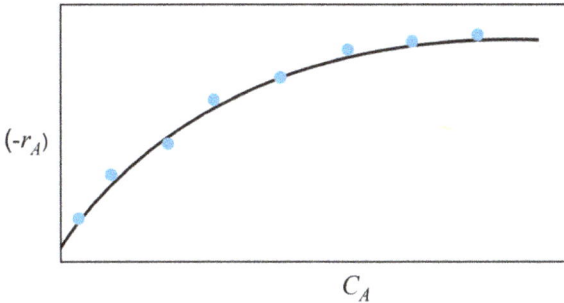

Figure 4.4: The behavior of eq. (4.20); C_B = constant.

Note that at low concentrations of C_A, the behavior is that of a first-order reaction, while at high concentrations, a zeroth-order reaction behavior is observed. As an exercise, examine the behavior of eq. (4.20) when C_B changes and C_A is maintained constant.

Laidler [5] reports an interesting example of a catalytic reaction following both the L-H and E-R mechanisms, when different catalysts are used. This is the ethylene hydrogenation reaction: under certain conditions, it follows the L-H mechanism over copper, whereas it follows the E-R mechanism over nickel.

4.4 Evaluation of kinetic parameters

In undergraduate courses on physical chemistry and reaction engineering, we typically work out the rate expressions for homogeneous reactions by *linearization*, so that we can determine the reaction order(s) and rate coefficient(s).

The same approach can be used in solid catalytic reactions. First, we start by proposing a set of elementary adsorption–reaction–desorption steps (consistent with the stoichiometric reaction) as we have done in the previous sections. Then a rate expression can be developed based on the RLS concept. When experimental rate data is available, we can then determine the "best" expression. This situation is demonstrated in Example 4.1.

When a satisfactory rate expression is derived, what remains for us is to evaluate the rate constant and adsorption equilibrium constants, which we normally call the *kinetic parameters*.

A relatively simple method of evaluating the kinetic parameters, in the case of L-H rate expressions, is to take the reciprocal of the expression, that is, invert the expression to arrive at a linear analysis.

This approach works for relatively simple rate equations. In general, however, **nonlinear regression analysis** [6] is used to discriminate among rival mathematical models. In recent years, other techniques have also been used [7].

Example 4.1:

The following reaction rate data were collected for the oxidation of hexafluoro-pentanedione in the presence of a Cu-based catalyst:

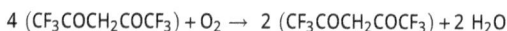

$$4 \ (CF_3COCH_2COCF_3) + O_2 \rightarrow 2 \ (CF_3COCH_2COCF_3) + 2 \ H_2O$$

Use the data below to check which of the two rate equations (adsorption without and with dissociation) better represents the kinetics of the reaction, where $A \equiv CF_3COCH_2COCF_3$. Note that P is expressed in torr and the rate in mol/g. min:

$$(-r_A) = \frac{kKP_{O_2}}{1 + KP_{O_2}}$$

$$(-r_A) = \frac{kK\sqrt{P_{O_2}}}{1 + K\sqrt{P_{O_2}}}$$

P_{O_2}	0.1	0.3	0.5	1.0	1.5	2.0	3.0	5.0	7.5	10.0	15.0	20.0
Rate	0.3	0.5	0.6	0.8	0.8	0.9	1.0	1.1	1.2	1.3	1.4	1.5

Solution: Let us convert both rate equations to a linear form so that we can test the experimental data. Taking the reciprocal of each equation, we obtain

$$\frac{1}{(-r_A)} = \left(\frac{1}{kK}\right)\frac{1}{P_{O_2}} + \left(\frac{1}{k}\right)$$

$$\frac{1}{(-r_A)} = \left(\frac{1}{kK}\right)\frac{1}{\sqrt{P_{O_2}}} + \left(\frac{1}{k}\right)$$

The plots below show that the second equation gives a better representation of the data.

Without Dissociation

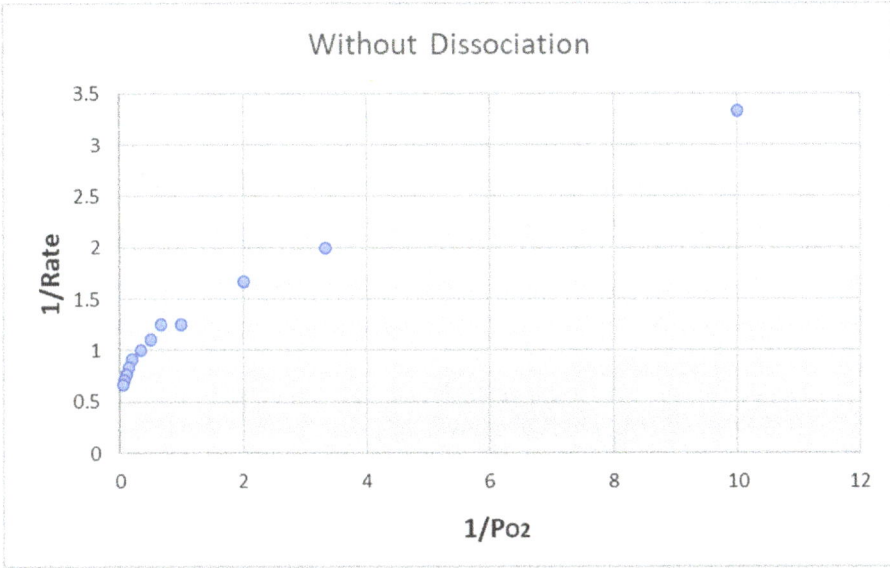

With Dissociation

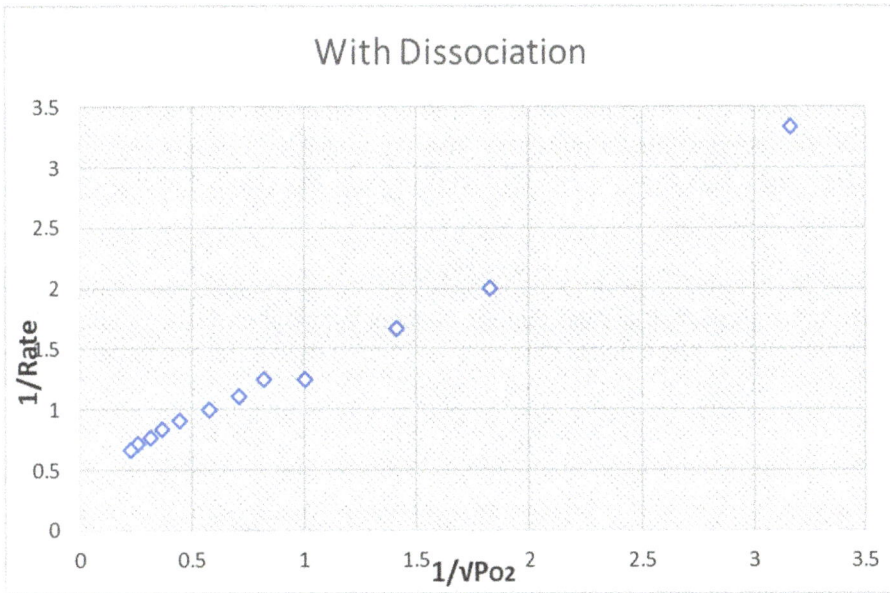

4.5 Effect of temperature on reaction rate

As we know the rate of a homogeneous reaction is strongly dependent on temperature. This strong effect is represented by the exponential dependence of the reaction rate coefficient on temperature, called the *Arrhenius law*.

Since solid-catalyzed reactions involve transport, adsorption, and surface reaction, we should naturally expect the rate of a heterogeneous reaction to be strongly dependent on temperature.

Indeed, the dependence is more complex. To illustrate some of the complexities involved, let us consider a simple heterogeneous reaction involving mainly the adsorption of the reactant species.

The rate of a simple irreversible unimolecular reaction in which the product is not adsorbed is given by eq. (4.4), which, in terms of the partial pressure of species A, becomes:

$$(-r_A) = k\theta_A = k \frac{K_A P_A}{1 + K_A P_A}$$

We can immediately see that the effect of temperature on the rate is the product of its effects on k and θ_A separately. Let us examine in detail what happens to the surface coverage of species A at limiting values of K.

For *large* values of K_A, $\theta_A \rightarrow 1$ throughout the examined temperature range; the effect of temperature is simply that on k. In this case, the **true** activation energy, E_{true}, can be obtained from the Arrhenius law, that is, $k = A\ e^{-E_{true}/RT}$.

For *small* values of K_A, we can see that $\theta_A \rightarrow K_A P_A$. As a result, the rate expression reduces to

$$(-r_A) = k\theta_A = kK_A P_A \tag{4.21}$$

From this expression, it is clear that the effect of temperature is the product of its effect on k and K_A. Recall that in Chapter 2 the effect of temperature on the adsorption constant is given by the van't Hoff relationship:

$$\frac{d \ln K_A}{dT} = \frac{\Delta H^o_{ads}}{RT^2} \tag{4.22}$$

where ΔH^o_{ads} is the *standard molar enthalpy of adsorption* of species A. Using eq. (4.22), we obtain

$$K_A = Ce^{-\frac{\Delta H^o_{ads}}{RT}} \tag{4.23}$$

Under these conditions, the rate expression becomes:

$$(-r_A) = \left(Ae^{-\frac{E_{true}}{RT}} \right) \left(Ce^{-\frac{\Delta H^o_{ads}}{RT}} \right) P_A$$

or

$$(-r_A) = A^* e^{-\frac{E_{app}}{RT}} \cdot P_A \tag{4.24}$$

where A^* equals the product (AC), and the **apparent** activation energy E_{app} is given by:

$$E_{app} = E_{true} + \Delta H_{ads}^\circ \tag{4.25}$$

From this, we immediately notice that

$$E_{true} = E_{app} - \Delta H_{ads}^\circ \tag{4.26}$$

Recall that since $\Delta H_{ads}^\circ < 0$, that is, adsorption is always exothermic, then we can alternatively write:

$$E_{true} = E_{app} + \left| \Delta H_{ads}^\circ \right| \tag{4.27}$$

This simple reaction demonstrates that the analysis of temperature effects in catalytic reactions is not as straightforward as in the case of homogeneous reactions. While an Arrhenius plot of the rate coefficient vs. temperature in such reactions would directly give a value of the activation energy, a similar plot for the reaction at hand could yield results similar to those shown in Figure 4.4.

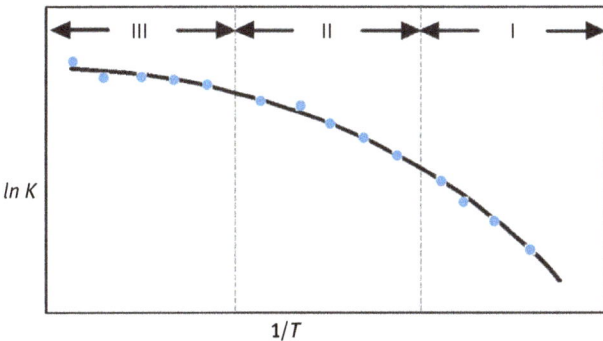

Figure 4.5: An Arrhenius plot for the simple catalytic reaction A → B.

Figure 4.5 shows what might be observed for *this* reaction over a wide temperature range. Here it is possible to identify three somewhat different regions, in each of which both the reaction order and the activation energy change, depending on the strength of the surface coverage. Table 4.1 summarizes the above results.

Table 4.1: The effect of change in temperature on the observed order and activation energy (adapted from [8]).

Region	θ	Order, n	Slope
I	~1	0	E_{true}/R
II	$\theta \in (0,1)$	$n \in (0,1)$	–
III	~ 0	1	E_{app}/R

What about **other** catalytic reactions? As was mentioned in the beginning of this section, the effect of temperature can be quite complex, and simple analyses as those done here are possible only for very few reactions.

The reader can consult Satterfield [9], for example, to see how the temperature affects the observed order, rate coefficient, and activation energy in the case of a simple irreversible, unimolecular, reaction in which both the reactant and product species are adsorbed.

Before we end this section, it should be mentioned that it is important to distinguish this cause of change in the reaction order and activation energy from a different cause that has a similar effect. We will study the effect of mass transport limitations on the observed kinetics in Chapter 5.

References

[1] Shaikh A. A., and Zaidi S. M. J. "Kinetics of Catalytic Oxidation of Aqueous Sodium Sulfite," React. Kinet. Catal. Lett, 64, 343 (1998).
[2] Mars P., and van Krevelen D. W. "Oxidations Carried Out by Means of Vanadium Oxide Catalysts," Chem. Eng. Sci., 3(suppl. 1), 41 (1954).
[3] Hougen O. A., and Watson K. M. "Chemical Process Principles, Part 3: Kinetics and Catalysis," Wiley, New York, 1947.
[4] Butt J. B. "Reaction Kinetics and Reactor Design," Second Ed., Marcel Dekker, New York, (2000).
[5] Laidler K. J. "Chemical Kinetics," Third Ed., Harper & Row, New York, (1987).
[6] Matera S., Schneider W. F., Heyden A., and Savara A. "Progress in Accurate Chemical Kinetic Modeling, Simulations, and Parameter Estimation for Heterogeneous Catalysis," ACS Catal., 9(8), 6624 (2019).
[7] Murzin D. Y., Wärnå J., Haario H., and Salmi T. "Parameter Estimation in Kinetic Models of Complex Heterogeneous Catalytic Reactions Using Bayesian Statistics," Reac. Kinet. Mech. Cat., 133, 1 (2021).
[8] Bond G. C. "Heterogeneous Catalysis: Principles and Applications," Second Ed., Oxford University Press, Oxford, (1987).
[9] Satterfield C. N. "Heterogeneous Catalysis in Industrial Practice," Second Ed., McGraw-Hill, New York, (1991).

5 Transport effects in catalytic reactions

Where did you get those values of the dimensionless groups (Biot, Hatta) . . . from the air!
A question (sarcastically) addressed to the author during a seminar.

We have briefly discussed the role of mass transport in the catalytic sequence in Chapter 1. We have also distinguished between two kinds of mass transport: *external* between the bulk fluid and surface of the solid particle; and *internal* within the typically porous structure of the solid particle. Note that "typically" porous means that the majority of industrial catalysts are porous.

The aim of this chapter is to closely examine the mass transport effects in catalytic reactions; quantifying these effects in terms of mass transfer rates; and discussing how catalytic reaction rates can be measured in the laboratory.

5.1 External and internal mass transfer

When considering a point in a catalytic reactor (recall Figure 1.7b) and focusing our attention on how reactant A reaches the surface and the internal parts of the solid catalyst, then several possibilities exist. Figure 5.1 shows the possibilities in a qualitative way in terms of the concentration profile *around* and *within* the solid particle.

The first possibility, case (a), shows that the concentration is uniform around and within the particle and equals that of species A in the bulk fluid. This implies that there is no resistance to mass transfer across the *film*[1] surrounding the solid particle, and no resistance within the particle itself. In case (b), there are significant mass transfer resistances around and within the particle.

In case (c), the mass transfer resistance is mainly located in the external film; whereas in case (d), the resistance is primarily located within the interior body of the particle.

Depending on the relative rates of the reaction, external transport, and internal transport, it is possible to say that a particular process is rate-controlling. The *general* case among (a)–(d) is (**b**), in which the rates of transport steps are comparable to the reaction rate. Case (c) corresponds to external-transport control; whereas case (d) corresponds to internal-transport control.

---

1 The student/reader should be familiar with the concept of a "film," that is, the fluid boundary layer surrounding the solid, from courses in transport phenomena. Likewise, the student/reader by now should be familiar with the concept of a "rate-limiting step," which we have used in Chapter 4.

https://doi.org/10.1515/9783111032511-005

(a)

External mass transfer film

(b)

(c)

(d)

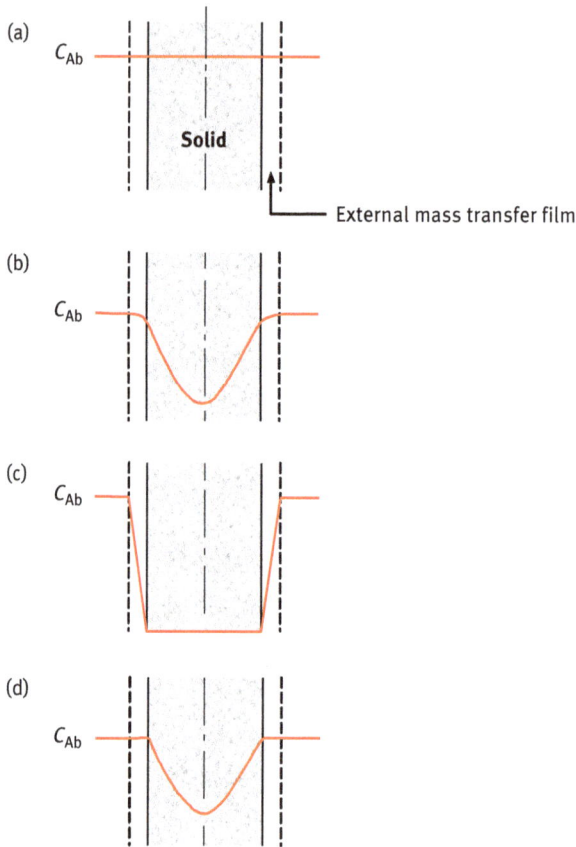

Figure 5.1: Concentration profiles inside and outside a catalyst particle: (a) reaction control, (b) general case, (c) external mass transfer control, and (d) internal mass transfer control.

5.2 The effectiveness factor

If we consider the general case of a catalytic reaction where concentration gradients exist outside and inside the solid; that is, when external and internal mass transfer resistances are significant, then it is clear that the availability of the reactant(s) near the active sites is limited. Consequently, the usefulness of the catalyst suffers.

An important measure of these transport limitations is the **effectiveness factor** that compares the actual reaction rate of the catalytic reaction to that in the absence of mass transfer limitations.

In order to carry out a quantitative analysis of the role of transport effects in heterogeneous catalysis, the case involving a first-order reaction is selected. It should however be kept in mind that the kinetics of heterogeneous reactions, as demonstrated in Chapter 4, are typically more complex.

Before proceeding to the mathematical description of transport–reaction interactions in catalyst particles, we should recall that a catalyst particle is generally not an *isotropic* medium (i.e., it does not have the same properties in all directions). This is due to two facts:

First, the catalyst particle is composed of *solid* materials (including the active agent(s), support, and promoters) as well as *void* spaces (which we have called pores). Second, the pores in the catalyst particle are not all of the same size and shape. As we have seen in Chapter 3, there is usually a distribution of pore sizes and the pores themselves are typically *tortuous*, that is, they are not straight. Figure 5.2(a) shows a schematic of a typical porous structure.

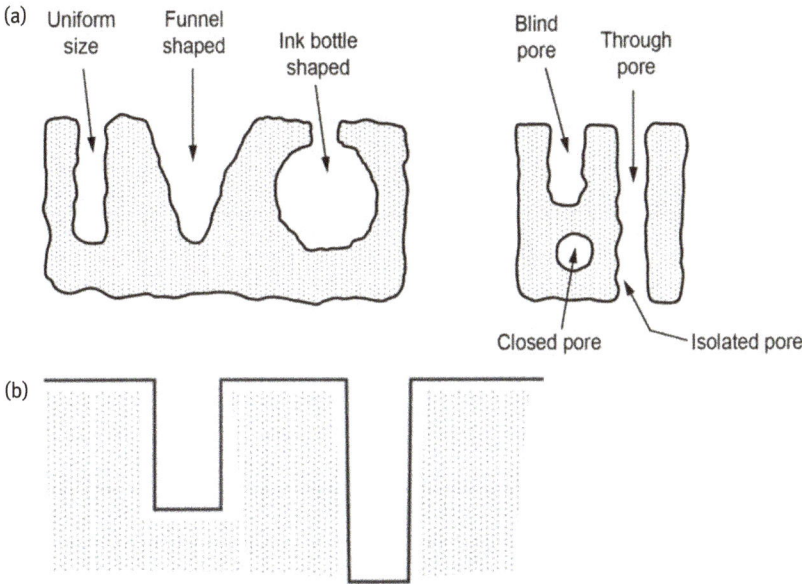

Figure 5.2: (a) A schematic of a typical porous network and pore shapes [1]. (b) A schematic of "ideal" pores.

These facts have led to the introduction of the concept of a *model particle*, in which the heterogeneity of the surface and the pores is taken into account by use of an *effective diffusion coefficient* (or effective diffusivity), and the assumption of an *ideal cylindrical pore* as shown in Figure 5.2(b). This coefficient, as will be seen later, is a function of the molecular diffusion coefficient, the porosity, and the tortuosity.

5.2.1 Isothermal first-order catalytic reactions

Consider a rectangular catalyst particle in which a first-order reaction $A \rightarrow B$ takes place under isothermal conditions. We shall consider the mass conservation of species A within the particle by applying the continuity equation [2]:

$$\frac{\partial C_A}{\partial t} + \frac{\partial N_{Ax}}{\partial x} + \frac{\partial N_{Ay}}{\partial y} + \frac{\partial N_{Az}}{\partial z} = R_A \tag{5.1}$$

where N_A is the total (convective and diffusive) flux and R_A is the $\pm$ generation term.

If the following assumptions are introduced:

(1) steady state,
(2) one-dimensional rectangular geometry,
(3) negligible convective flux, and
(4) constant effective diffusivity.

Equation (5.1) reduces to the following simple form for reactant A:

$$D_e \frac{d^2 C_A}{dx^2} = (-r_A) \tag{5.2}$$

where $(-r_A)$ is the rate of consumption of species A by the chemical reaction. In the present case $(-r_A) = kC_A$.

The boundary conditions for eq. (5.2) can be written by referring to Figure 5.1, case (**b**), where there is symmetry at the center of the particle and mass transfer resistance at the surface; mathematically we get:

$$x = 0, \quad \frac{dC_A}{dx} = 0 \tag{5.3}$$

$$x = +L, \quad \left(-D_e \frac{dC_A}{dx} \right) = -k_g(C_{Ab} - C_A) \tag{5.4}$$

where k_g is the external mass transfer coefficient; C_{Ab} is the bulk-phase concentration of species A; and L is the half-thickness of the particle. It should be noted that bulk-phase concentration refers to the concentration external to the particle at a point in the reactor.

Equations (5.2)–(5.4) can be rewritten in dimensionless form by defining

$$a \equiv \frac{C_A}{C_{Ab}} \text{ where } a \in [0, 1]; \text{ and}$$

$$z \equiv \frac{x}{L} \quad \text{where } z \in [0, 1]$$

Therefore, eqs. (5.2)–(5.4) reduce to the following boundary-value problem:

$$\frac{d^2a}{dz^2} - \phi^2 a = 0 \tag{5.5}$$

$$z = 0, \quad \frac{da}{dz} = 0 \tag{5.6}$$

$$z = 1, \quad \frac{da}{dz} = Bi(1 - a) \tag{5.7}$$

The dimensionless formulation of the problem has led to two groups: ϕ called the **Thiele modulus**;[2] and Bi called the **Biot number**. Both dimensionless groups have special significance in catalysis.

In our problem, the Thiele modulus is defined as follows:

$$\phi^2 = \frac{kL^2}{D_e} \equiv \frac{\text{reaction rate}}{\text{internal mass transfer rate}} \tag{5.8}$$

A value of $\phi \gg 1$ thus implies that the catalytic reaction is diffusion-controlled; whereas a value of $\phi \ll 1$ implies that the catalytic reaction is chemical reaction-controlled (or kinetics-controlled).

The Biot number is defined as follows:

$$Bi = \frac{k_g L}{D_e} \equiv \frac{\text{external mass tranfer rate}}{\text{internal mass transfer rate}} \tag{5.9}$$

Therefore, a value of $Bi \gg 1$ implies that $C_A(\pm L) \to C_{Ab}$, so that the catalytic reaction is *either* diffusion-controlled *or* kinetics-controlled; whereas $Bi \ll 1$ *could* indicate that the catalytic reaction is external mass-transfer-controlled.

Equation (5.5) is a linear, second-order, ordinary differential equation with constant coefficients, and can thus be solved analytically. The general solution is

$$a = C_1 \sinh(\phi z) + C_2 \cosh(\phi z) \tag{5.10}$$

By using boundary conditions (5.6) and (5.7) we can determine the integration constants, C_1 and C_2, so that the solution becomes:

$$a(z) = \frac{\cosh(\phi z)}{\cosh(\phi)\left[1 + \frac{\phi}{Bi}\tanh(\phi)\right]} \tag{5.11}$$

Equation (5.11) gives us the concentration profile within the catalyst particle. Figure 5.3 shows typical profiles, where $z = 0$ represents the particle center and $z = 1$ represents the *right-hand side* of the particle.

---

2 This is indeed the same Thiele whose name appears in the McCabe–Thiele diagram. E.W. Thiele is considered to be one of the founding fathers of chemical engineering science.

The reader should study these curves carefully to see the effect of mass transfer within the particle (when ϕ changes), and outside the particle (when Bi changes). In general, these effects lead to a decrease in the global (observed) reaction rate because of the reduction in reactant concentration across the film from the external surface to the center of the particle.

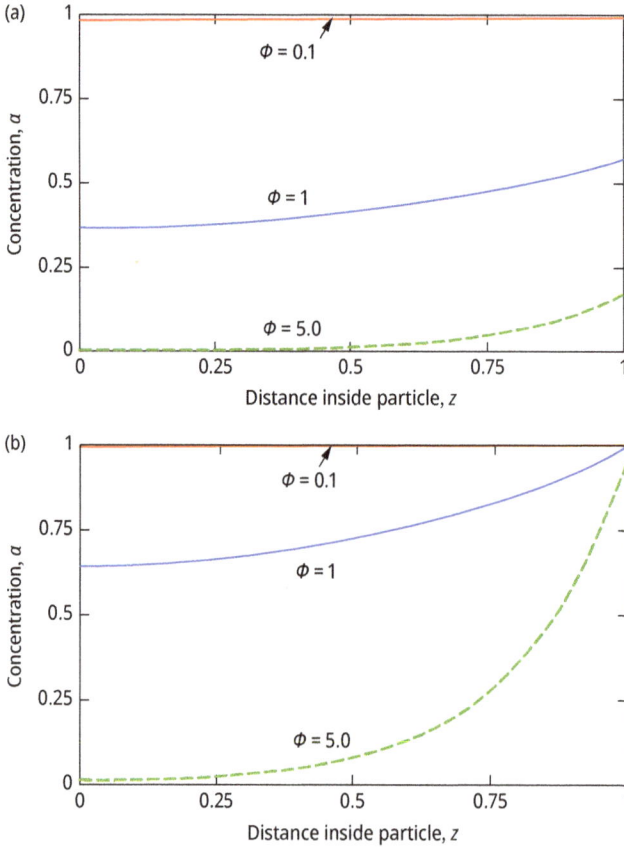

Figure 5.3: (a) Concentration profiles within slab-like catalyst particle; $Bi = 1$. (b) Concentration profiles within slab-like catalyst particle; $Bi = 100$.

5.2.2 Effectiveness factor for first-order reactions

As already mentioned, an important index for quantifying how both internal and external mass transfer limit the rate of catalytic reaction is represented by the effectiveness factor, η, which is defined[3] according to:

$$\eta = \frac{(-r_A)}{(-r_A)_b} = \frac{\text{actual rate of reaction}}{\text{rate of reaction without mass transfer effects}} \tag{5.12}$$

Note that the reaction rate without mass transfer effects is evaluated at C_{Ab}. This is an *observable* variable, that is, a quantity that is measurable in the laboratory or industrial reactor, as opposed to C_A within the solid, which is generally not an observable variable. Thus, once a value of η is obtained, the actual rate of the reaction can be directly determined.

Based on the definition of the effectiveness factor, η can be expressed as follows:

$$\eta = \frac{\frac{1}{L}\int_0^L (-r_A)dx}{(-r_A)_b} \tag{5.13}$$

In the case of first-order kinetics, eq. (5.13) reduces to the simple integral:

$$\eta = \int_0^1 a\,dz \tag{5.14}$$

Equations (5.11) and (5.14) can now be combined to yield the following expression for the effectiveness factor of a slab-like catalyst particle:

$$\eta = \frac{\tanh(\phi)}{\phi\left[1 + \frac{\phi}{Bi}\tanh(\phi)\right]} \tag{5.15}$$

Figure 5.4 shows the behavior of the effectiveness factor as a function of the Thiele modulus and Biot number. We observe that the effectiveness of the catalyst is reduced in the case of significant diffusion resistance and also in the presence of severe external mass transfer resistance

It is instructive to consider the asymptotic limits of eq. (5.15) when the external mass transfer effects are negligible, that is, when $Bi \rightarrow \infty$. In that limit, the effectiveness factor reduces to the following simple expression:

---

3 Strictly speaking, this definition is somewhat incomplete as it excludes the possible **heat transfer effects** around and within the particle; however, the consideration of such effects is beyond the scope of this chapter.

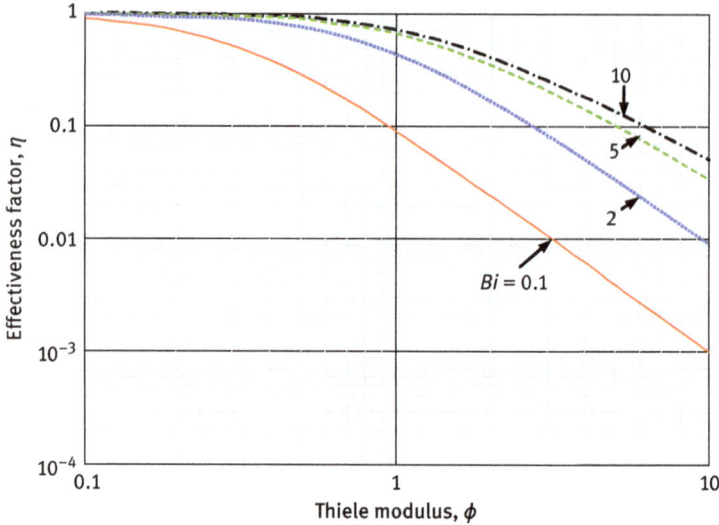

Figure 5.4: The effectiveness factor as a function of the Thiele modulus and Biot number for a first-order reaction in a rectangular (slab-like) catalyst particle.

$$\eta = \frac{\tanh(\phi)}{\phi} \tag{5.16}$$

which, in turn, reduces to the following limits at large and small values of Thiele modulus:

$\eta \rightarrow 1$, when $\phi \ll 1$: the reaction is said to be *kinetics-controlled*; and

$\eta \rightarrow \frac{1}{\phi}$, when $\phi \gg 1$: the reaction is said to be severely *pore-diffusion limited*.

These limits are also reflected in Figure 5.4. We conclude therefore that pore diffusion, that is, internal mass transfer, can severely affect the rate of catalytic reactions.

5.2.3 Effectiveness factors for other catalyst shapes

The analysis of the effectiveness factor presented in the previous section can be extended to other, more common, catalyst shapes. More specifically we will consider results for η in the case of first-order reactions, and in the case of finite diffusional resistance and finite external mass transfer resistance. Table 5.1 summarizes the results.

Recall that ϕ_0 is given by eq. (5.8); whereas the Thiele moduli for cylindrical and spherical particles are defined by $\phi_{1,2} = \sqrt{\frac{kR^2}{D_e}}$, where R is the radius of the cylinder or sphere.

Table 5.1: The effectiveness factors for particles of different shapes; first-order reaction.

Case	Shape	η	$\eta \ (\phi \rightarrow \text{large})$
0	Slab	$\dfrac{\tanh(\phi)}{\phi + \left(\frac{\phi^2}{Bi}\right)\tanh(\phi)}$	$\dfrac{1}{\phi_0}$
1	Cylinder[a]	$\dfrac{2\,I_1\,(\phi)}{\phi\left[I_0\,(\phi) + \left(\frac{\phi}{Bi}\right)I_1(\phi)\right]}$	$\dfrac{2}{\phi_1}$
2	Sphere	$\dfrac{3}{\phi^2}\left[\dfrac{\phi\coth(\phi) - 1}{1 + \frac{1}{Bi}\,(\phi\coth(\phi) - 1)}\right]$	$\dfrac{3}{\phi_2}$

[a]$I_0(\phi_1)$ and $I_1(\phi_1)$ are modified Bessel functions of the first kind of orders zero and one. Such functions are called "special" and they are easily evaluated by MS Excel® and other computational software.

Example 5.1 (modified data based on [3]): A commercial process for the dehydrogenation of ethyl benzene employs 3 mm spherical catalyst particles. The reaction rate constant is 15 s^{-1}, and the diffusivity of ethyl benzene in steam is 4×10^{-5} m^2/s under reaction conditions.
(a) Write down the chemical equations for this reaction.
(b) Explain the industrial importance of this reaction.
(c) Determine the isothermal effectiveness factor.

Solution:
(a) $C_6H_5CH_2CH_3 \xrightarrow{\text{excess steam}} C_6H_5CH = CH_2 + H_2$
(b) The main product is styrene which is the monomer required to produce polystyrene. Polystyrene is one of the most commercially produced plastics.
(c) We can use the expression in Table 5.1 for spherical catalysts where $R = 1.5$ mm:

$$\phi = \sqrt{\frac{kR^2}{D_e}} = 0.92$$

Assuming external mass transfer is negligible as no information is given,

$$\eta = \frac{3}{\phi^2}[\phi\coth(\phi) - 1] = 0.75$$

5.2.4 Parameter estimation

We have shown that the actual rate of the catalytic reaction is a function of the effectiveness factor. This factor is in turn a function of the Thiele modulus and Biot number. To estimate the effectiveness factor we need to calculate the mass transfer coefficient and the effective diffusivity. In short we can say that

$$-r_A = \eta(-r_A)_b \tag{5.17}$$

where $\eta = f(\phi, Bi) = f(k, k_g, D_e)$.

In the last section of this chapter we will briefly address experimental methods for the determination of the "true" or "**intrinsic**" (i.e., unaffected by mass transfer) reaction rate constant. However, here we shall consider ways for evaluating the mass transfer coefficient and effective diffusivity.

a. Mass transfer coefficient (k_g)

From our background in transport phenomena, we know that empirical correlations of the form $Sh = f(Re, Sc)$ can be used to estimate the external mass transfer coefficient, where Sh, Re, and Sc are the Sherwood, Reynolds, and Schmidt numbers, respectively. An example is the Frössling correlation [4], for mass transport to single solid spheres:

$$Sh = 2 + 0.6\ Re^{\frac{1}{2}} Sc^{\frac{1}{3}} \tag{5.18}$$

where $Sh = k_g d/D_{AB}$, $Re = u_\infty \rho d/\mu$, and $Sc = \mu/\rho D_{AB}$. Here, u_∞ is the "free-stream" velocity, and d is particle diameter.

Another correlation for flow over solid spheres is given by the following expression [5]:

$$Sh = 2 + 0.4\left(Re^{\frac{1}{2}} + Re^{\frac{2}{3}}\right) Sc^{0.4} \tag{5.19}$$

b. Effective diffusivity (D_e)

The consideration of mass transfer by diffusion in *porous* catalyst particles is somewhat more complicated compared to ordinary molecular diffusion that we are familiar with from courses in mass transfer.

As was demonstrated in Chapter 3, a distribution of pore sizes in general exists within the particle, and the pores are tortuous in nature. Figure 5.5 illustrates two types of pores compared to the so-called *mean free path* for molecular interactions (δ). We should recall that according to the Kinetic Theory of Gases, the mean free path is defined as the average distance a molecule travels between collisions.

When $d_p \gg \delta$, ordinary **bulk diffusion** occurs, which we characterize by D_{AB}, the ordinary molecular diffusivity of A in the binary system (A + B). Most of us probably still remember that the molecular diffusivity is evaluated using the following expression by Hirschfelder et al. [6]:

$$D_{AB} = \frac{0.00186T^{\frac{3}{2}}\left[\frac{1}{M_A} + \frac{1}{M_B}\right]^{\frac{1}{2}}}{\sigma_{AB}^2 \Omega_D P} \tag{5.20}$$

where the units of D_{AB} are (cm^2/s); T is the temperature in (K); M_A and M_B are the molecular weights of A and B, respectively; P is pressure in (atm); σ_{AB} is the Lennard–Jones collision diameter in (Å); and Ω_D is the so-called collision integral for molecular diffusion, which is dimensionless.

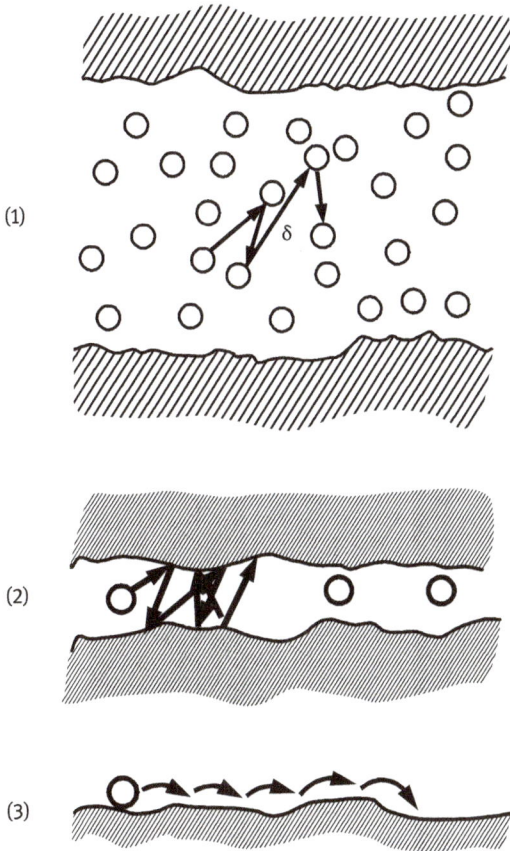

Figure 5.5: Diffusion in a porous catalyst particle. (1) Bulk diffusion; (2) Knudsen diffusion; and (3) Surface diffusion.

When $d_p < \delta$, **Knudsen diffusion** occurs, where collision of the molecules with the pore walls is dominant. This mode of diffusion is characterized by D_K, where

$$D_K = 9700 \ \bar{r}_P \sqrt{\frac{T}{M_A}} \tag{5.21}$$

where D_K is expressed in (cm^2/s); $\bar{r}_P$ is the average pore radius (cm); T is the temperature (K), and M_A is the molecular weight of the reactant.

The effective diffusivity can now be evaluated based on the above expressions for bulk and Knudsen diffusion according to:

$$D_e = \frac{\varepsilon_p}{\tau_p} \left[\frac{1}{\frac{1}{D_{AB}} + \frac{1}{D_K}} \right] \tag{5.22}$$

where ε_p is the *particle porosity* and τ_p is the *geometric tortuosity factor*. These are necessary correction factors to account for the facts that the pores are not straight and the flow of the diffusing molecules is restricted by the solid fraction of the catalyst particle.

Table 5.2 lists experimental values for ε_p and τ_p which have been reported in the literature for various solids.

Table 5.2: Porosity and tortuosity factors for diffusion in catalysts (adapted from [7, 8]).

Catalyst	ε_p	τ_p
Pelletized Cr_2O_3/Al_2O_3	0.22	2.5
Pelletized bohemite/Al_2O_3	0.34	2.7
Girdler G-58 Pd/Al_2O_3	0.39	2.8
Haldor Topsoe methanol synthesis catalyst	0.43	3.3
BASF methanol synthesis catalyst	0.50	7.5
0.5% Pd/Al_2O_3	0.59	3.9
1.0% Pd/Al_2O_3	0.50	7.5
Pelletized Ag/8.5% Ca alloy	0.30	6.0
Pelletized Ag	0.30	10.0

Figure 5.5 also shows a third kind of diffusion within the pores, which is called **surface diffusion**. This mode of diffusion results from detachment and migration of molecules along the pore surface; this type of diffusion may or may not be important. Guidelines as to how one should account for surface diffusion when estimating the effective diffusivity have been proposed in the literature. However, we concern ourselves here with the first two modes of diffusion, as a discussion of the third mode is beyond the scope of this chapter.

Example 5.2 (adapted from [9]): Estimate the Knudsen diffusivity and the effective diffusivity of thiophene (C_4H_4S) at 660 K and 30 bar in a catalyst particle with an inside surface area of 180 m^2/g, a porosity of 0.4, and a particle density of 1,400 kg/m^3. The tortuosity factor is 2. It is also known that the concentration of thiophene is very low and the pore fluid consists mainly of hydrogen. Under the above conditions the bulk diffusivity of thiophene in hydrogen is 5.2 × 10^{-6} m^2/s.

Solution:

$$D_K = 9700 \; \overline{r}_p \sqrt{\frac{T}{M_A}} \qquad D_e = \frac{\varepsilon_p}{\tau_p}\left[\frac{1}{\frac{1}{D_{AB}} + \frac{1}{D_K}}\right]$$

$$M_A = 84$$

According to eq. (3.3),

$$\overline{r}_p = \frac{2V_g}{S_g}$$

$$\varepsilon_p \equiv \frac{V_{pore}}{V_{particle}} = \frac{V_g \times mass}{V_{particle}} = \frac{V_g}{V_{particle}/mass} = \frac{V_g}{(1/\rho_{particle})}$$

$$V_g = \frac{\varepsilon_p}{\rho_p} = \frac{0.4}{1,400 \times 10^3} = 2.86 \times 10^{-7} m^3/g$$

$$\overline{r}_p = 3.2 \, nm \Rightarrow D_K = 8.6 \; 10^{-7} \, m^2/s; \; and \; D_e = 1.47 \times 10^{-7} \, m^2/s$$

5.3 Behavior of observed reaction rate: falsification of kinetic data

Now that we have seen how internal and external mass transfer can limit the rate of catalytic reactions, it is important to understand how these phenomena affect the behavior of the observed (measured) reaction rate.

This section is devoted to how mass-transfer effects can cause *difficulties in the interpretation of kinetic data*. This problem has been termed "falsification" of kinetic data in catalytic reactions.[4]

For the sake of illustration and mathematical simplicity of this subject, we shall consider the case involving a first-order reaction. We shall now examine in some detail the behavior of eq. (5.15) in three limiting cases.

Case 1. Chemical reaction control: ϕ = small; Bi = large

$$\eta \to 1 \Rightarrow (-r_A)_{obs} = kC_{Ab} \therefore k_{obs} = k_{true}$$

Case 2. Internal diffusion control: ϕ = large; Bi = large, such that $\dfrac{\phi}{Bi} \ll 1$

$$\eta \to \frac{1}{\phi} \Rightarrow (-r_A)_{obs} = \sqrt{\frac{D_e}{kL^2}}(kC_{Ab}) \therefore k_{obs} = \sqrt{k_{true}}$$

Case 3. External mass transfer control: ϕ = large; Bi = finite, such that $\dfrac{\phi}{Bi} \gg 1$

$$\eta \to \frac{Bi}{\phi^2} \Rightarrow (-r_A)_{obs} = \frac{k_g}{kL}(kC_{Ab}) \therefore k_{obs} = \frac{k_g}{L}; \text{ which means } k_{obs} \, \alpha \, k_g$$

---

4 Other terms such as "masking," "disguise," and "camouflage" have also been used.

The three cases involve significant differences, and it is *possible* to see results of reaction kinetic studies as summarized in Table 5.3. As an exercise, the student should try to see how the results in Table 5.3 were arrived at, especially the dependence on the particle size.

Table 5.3: A summary of the characteristics of internal and external mass transfer effects.

Regime control	Observed E	Effect of particle size*
Chemical reaction	E_{true}	None
Internal diffusion	$(E_{true})/2$	$(1/L)$
External mass transfer	$\rightarrow 0$	$(1/L)^a$

*The exponent "a" is a function of the catalyst shape and reactor type.

To summarize the results in this section, we conclude by saying that **the intrinsic rate constant is not dependent on the scale and reactor, but the associated physical effects (mass and heat transfer) are dependent on the scale and reactor**. Therefore, the physical effects can seriously affect the rate of the catalytic reaction.

5.4 Laboratory reactors

Based on the results of the previous section, one can say that the determination of the true (or intrinsic) kinetics, as opposed to the "falsified" kinetics, of solid-catalyzed reactions requires special care.

In fact, many experimental techniques, quite different from those that we are familiar with from our background in *homogenous* reaction kinetics (e.g., the batch reactor), have been developed and used over the years. Table 5.4 provides descriptions of some of the laboratory reactors that have been developed.

Table 5.4: Some examples of experimental bench-scale reactors for catalytic reactions (more details are available in [10]).

Name	Category	Typical description
Differential reactor	Fixed-bed tubular reactor	0.5 cm diameter tubular reactor with a very small amount of catalyst
Integral reactor	Fixed-bed tubular reactor	2.5 cm diameter, around 30 cm long tubular reactor
Carberry reactor	Mixed-flow reactor	Spinning catalyst-basket CSTR
Berty reactor	Mixed-flow reactor	CSTR where flowing gas is recirculated through catalyst by means of an impeller

When collecting catalytic reaction rate data in laboratory reactors, one should ensure that the following conditions are satisfied:[5]
(a) An isothermal environment is maintained.
(b) The effect of internal and external transport is minimized. Laboratory reactors satisfying this condition are called *"gradientless"* reactors.

While it is easy to imagine how condition (a) can be satisfied, many efforts over the years have been expended to design and apply gradientless reactors. (Do you think the "Integral reactor" listed in Table 5.4 is gradientless?)

Let us now focus on one of these reactors, namely the **Carberry** reactor [11, 12], with the aim of studying its characteristics and how to use it to collect rate data. Although the original design of the Carberry reactor has been improved over the years, we discuss its operation here because of its historical value.

A sketch of this reactor is shown in Figure 5.6. The vigorous spinning of the mesh baskets, shown in Figure 5.7, containing the catalyst particles leads to increase in tur-

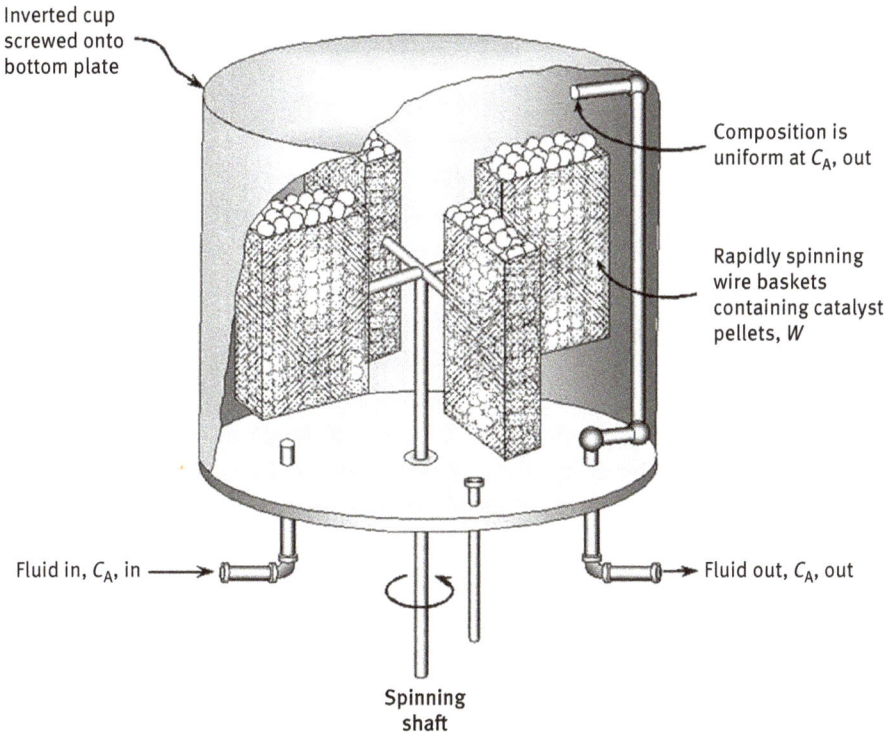

Inverted cup screwed onto bottom plate

Composition is uniform at C_A, out

Rapidly spinning wire baskets containing catalyst pellets, W

Fluid in, C_A, in

Fluid out, C_A, out

Spinning shaft

Figure 5.6: The experimental Carberry spinning-basket mixed flow reactor (adapted from [13]).

5 Another important requirement is that catalyst *deactivation* is considered if it is significant. We will take up the issue of deactivation in the next chapter.

bulence around the particles such that the *external* mass transfer resistance is minimized (recall eq. (5.18)).

The *internal* mass transfer resistance, on the other hand, is minimized by loading relatively small particles in the baskets (recall Table 5.3). The continuous spinning of the baskets allows us to assume that a well-mixed isothermal environment is maintained.

Therefore, the performance equation for a CSTR (the Carberry reactor in our case) as applied to a catalytic reaction involving species A is given as follows:

$$F_{A0} = F_A + (-r'_A)W \tag{5.23}$$

where F_{A0} is the feed molar flow rate, $(-r'_A)$ is the catalytic reaction rate in (mol/g catalyst-time), and W is the mass of the catalyst used in (g). In terms of the conversion X_A and molar flow rate F_{A0}, eq. (5.23) becomes:

$$(-r'_A) = \frac{F_{A0}X_A}{W} \tag{5.24}$$

Each experimental run (say, by changing F_{A0}) would, therefore, directly give us a value for the reaction rate at the composition of the exit fluid.

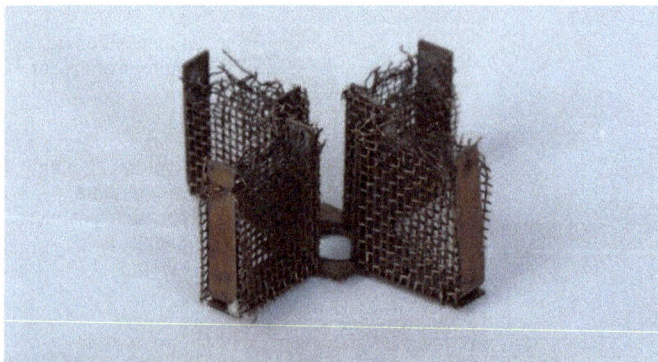

Figure 5.7: A photograph of old wire-mesh baskets in the Carberry reactor.

Example 5.3 (data adapted from [13]): The following kinetic data were obtained in an experimental Carberry-type reactor using 100 g of catalyst in the baskets and various flow rates, using $C_{A0} = 10$ mol/m^3 in all cases. Find a rate expression given that the reaction chemistry is A → R.

$F_{A,in}$ [mol/min]	0.14	0.42	1.67	2.5	1.25
$C_{A,out}$ [mol/m^3]	8	6	4	2	1

Solution:

We can use eq. (5.24) to find values of $(-r_A')$ at different values of $C_{A,out}$ using:

$$X_A = 1 - \frac{C_{A,out}}{C_{A0}}$$

Using the given data, we can construct the following table:

X_A	0.2	0.4	0.6	0.8	0.9
$(-r_A')$	0.0003	0.0017	0.0100	0.0200	0.0113

Before we assume a rate from, let us plot the reaction rate versus concentration to get an idea about the behavior of the rate equation, according to what we have learned in Chapter 4.

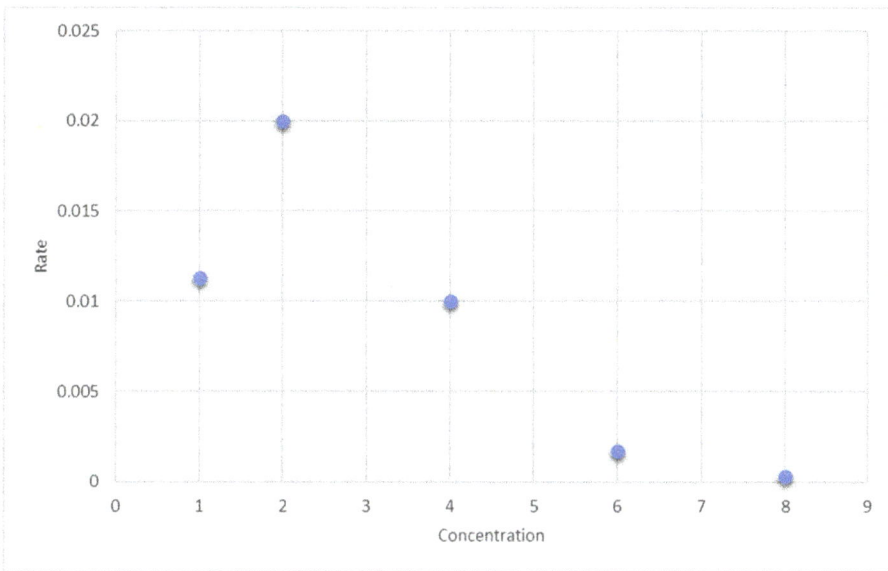

It is possible to conclude the following:
(1) nth-order kinetics are not shown by the given data.
(2) Based on similar curves in Chapter 4, it is possible that the reaction follows an L-H mechanism. But we do not have enough information to suggest a mechanism and a rate equation.

References

[1]   Leofanti G., Padovan M., Tozzola G., and Venturelli B. "Surface Area and Pore Texture of Catalysts," Catal. Today, 41, 207 (1998).

[2]   Welty J. R., Wicks C. E., Wilson R. E., and Rorrer G. L. "Fundamentals of Momentum, Heat, and Mass Transfer," Fifth Ed., Wiley, Chichester, (2008).

[3]   Wu J. C. "Dehydrogenation of Ethylbenzene over TiO_2-Fe_2O_3 and ZrO_2-Fe_2O_3 Mixed Oxide Catalysts," Catal. Lett., 20, 191 (1993).

[4]   Fogler H. S. "Elements of Chemical Reaction Engineering," Fifth Ed., Prentice Hall, Boston, MA, (2016).

[5]   Schmidt L. D. "The Engineering of Chemical Reactions," Second Ed., Oxford University Press, New York, (2005).

[6]   Masel R. I. "Chemical Kinetics and Catalysis," Wiley-Interscience, New York, (2001).

[7]   Rawlings J. B., and Ekerdt J. G. "Chemical Reactor Analysis and Design Fundamentals," Second Ed., Nob Hill Publishing, Santa Barbara, CA, (2020).

[8]   Butt J. B. "Reaction Kinetics and Reactor Design," Second Ed, Marcel Dekker, New York, (2000).

[9]   Winterbottom J. M., and King M. B. eds., "Reactor Design for Chemical Engineers", Stanley Thomas Publishers, Cheltenham, U.K., (1999).

[10] Bartholomew C. H., and Farrauto R. J. "Fundamentals of Industrial Catalytic Processes," Second Ed., Wiley, New York, (2006).

[11] Tajbl D. G., Simons J. B., and Carberry J. J. "Heterogeneous Catalysis in Continuous Stirred Tank Reactor," Ind. Eng. Chem. Fundam., 5, 171 (1966).

[12] Carberry J. J. "Chemical and Catalytic Reaction Engineering," McGraw-Hill, New York, (1976), Reprint edition by Dover Publications, New York, (2001).

[13] Levenspiel O. "Chemical Reaction Engineering," Third Ed., Wiley, New York, (1999).

6 Deactivation of solid catalysts

No talent goes unpunished.
Mohammad Al-Maghout

It was occasionally mentioned in previous chapters that solid catalysts deactivate. Deactivation simply means *loss of catalyst activity with usage*.

In industrial applications, the deactivation behavior of catalysts is as important as its activity or selectivity characteristics. Indeed, the deactivation of solid catalysts can be a major financial burden in the chemical and refining process industries.

The process engineer should understand how deactivation happens, as this understanding is the key to counter-balancing, at least partially, its effect on the production requirements, or to slowing down this unavoidable fact.

We, therefore, speak of catalyst *lifetime*, which may be in the order of a *few years* or, believe it or not, a *few seconds*! Table 6.1 provides examples of typical catalyst lifetimes for various industrial catalytic processes.

The objective of this chapter is to define the causes and mechanisms of catalyst deactivation. We shall also briefly consider methods for *regeneration* of spent catalysts, that is, methods by which the catalyst activity can be retained, (almost) fully restored, or partially restored.

Table 6.1: Examples of typical lifetimes of various industrial catalytic processes (adapted from [1, 2]).

Reaction	Operating conditions	Catalyst	Typical life (year)	Cause(s) of deactivation
Ammonia synthesis	450–550 °C 200–500 atm	$Fe/(K_2O)/(Al_2O_3)$	5–10	Sintering
Methanization (ammonia and hydrogen plants)	250–350 °C 30 atm	$Ni/(Al_2O_3)$	5–10	Poisoning (by S and As compounds)
Acetylene hydrogenation ("front end")	30–100 °C 50 atm	Supported Pd	5–10	Sintering
Sulfuric acid	420–600 °C 1 atm	V and K sulfates/ SiO_2	5–10	Attrition
Methanol synthesis (low temperature)	200–300 °C 50–100 atm	$Cu/Zn/(Al_2O_3)$	2–8	Sintering
Hydrodesulfurization of light petroleum cuts	300–400 °C 30 atm	CoS and $Mo_2S/$ (Al_2O_3)	0.5–2	Slow coking
Low-temperature carbon monoxide shift	200–250 °C 30 atm	Co and Zn and A1 oxides	2–6	Poisoning and sintering

https://doi.org/10.1515/9783111032511-006

Table 6.1 (continued)

Reaction	Operating conditions	Catalyst	Typical life (year)	Cause(s) of deactivation
High-temperature carbon monoxide shift	350–500 °C 30 atm	Fe_3O_4/chromia	2–4	Sintering and pellet breakage
Steam reforming of natural gas	500–850 °C 30 atm	Ni/calcium aluminate or α-alumina	2–4	Sintering, poisoning (S), carbon formation, and pellet breakage
Partial oxidation of ethylene to ethylene oxide	200–270 °C 10–20 atm	Ag/α-alumina	1–4	Sintering
Benzene oxidation to maleic anhydride	350 °C 1 atm	V_2O_5 and MoO_2/ (Al_2O_3)	1–2	Irreversible formation of inactive V phase
Reduction of aldehydes to alcohols	220–270 °C 100–300 atm	Co/Zn oxide	0.5–1	Sintering, and pellet breakage
Partial oxidation of methanol to formaldehyde	600–700 °C 1 atm	Silver granules	0.3–1	Poisoning (Fe) and coking
Oxychlorination of ethylene to ethylene dichloride	230–270 °C 1–10 atm	Cu chlorides/ (Al_2O_3)	0.2–0.5	Attrition
Ammonia oxidation	800–900 °C 1–10 atm	Pt-alloy gauze	0.1–0.5	Vapor loss of Pt and poisons
Acetylene hydrogenation ("tail end")	30–100 °C 50 atm	Supported Pd	0.1–0.5	Coking
Hydrocarbon reforming	460–525 °C 8–50 atm	Pt alloys/(Al_2O_3)	0.01–0.5	Coking
Fluid catalytic cracking (FCC) of gas oil	500–600 2–3 atm	Zeolites	0.0000001	Rapid coking

Data in the table are rather old; the table is used here mainly for illustrative purposes.

6.1 Causes of deactivation

You may recall that the concept of *active sites* on catalytic surfaces was introduced in Chapter 1. Catalyst deactivation can be viewed as "the overall result of the removal of active sites from the surface" [3].

This result happens by different reasons. In general, there are five major causes for catalyst deactivation:

(a) thermal degradation (or sintering);
(b) fouling (or coking);
(c) poisoning;
(d) evaporation (or volatilization) of some active catalyst agent(s);[1] and
(e) loss of mechanical strength of the solid particles by attrition or crushing.

It is instructive to consider the classifications and subclassifications of these mechanisms presented in Figure 6.1. The term "temporal" means a decline in catalyst activity with the **passage of time**. "Spatial" mechanisms mean loss of activity at **specific locations** on the catalyst surface. Furthermore, Figure 6.1 shows that these mechanisms can be further classified as either *chemical, physical*, or *mechanical*.

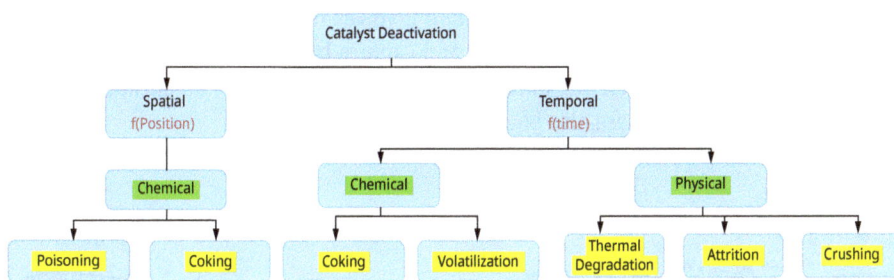

Figure 6.1: A broad classification of mechanisms of catalyst deactivation (adapted and modified from [5]).

Poisoning and evaporation (or volatilization) are essentially chemical in nature, whereas fouling, attrition, and crushing are physical (or mechanical) phenomena. We should be careful **not** to use this figure to draw strict boundaries between the deactivation mechanisms.

Likewise, we should note that the deactivation mechanisms can occur either *individually* or *in combination*, which means that more than one cause could be responsible for the deactivation in specific applications. Examples of mixed causes are given in Table 6.1.

The discussion in this chapter covers mainly the three mechanisms that are probably the most studied and thus the most important: thermal degradation, fouling, and poisoning. A brief look at the last two mechanisms is also presented.

---

1 According to [4], this cause is divided into (a) vapor formation by reaction of fluid with catalyst phase and (b) vapor-solid and solid-solid reactions.

6.2 Deactivation by thermal degradation

Sintering and *ageing* are other names that have also been used to designate this mechanism. As shown in Table 6.1, catalytic processes often operate under relatively high temperatures, such that the catalyst is gradually degraded.

Thermal degradation leads to *structural changes* in the catalytic surface and thus to a loss of active sites. This mechanism of deactivation is therefore essentially physical in nature.

Typical structural changes, in supported metal catalysts, may result from migration followed by agglomeration of microscale active agent particles (e.g., tiny metal crystallites) into larger ones, as well as atomic migration. As shown in Figure 6.2, crystallite migration may involve the migration of entire crystallites over the support surface followed by collision and coalescence. Atomic migration involves the detachment of metal atoms from crystallites, migration of these atoms, and capture by larger crystallites [6]. In supported catalysts, an actual collapse of the internal pore structure and an encapsulation of the active catalytic agent(s) are also possible, as shown in Figure 6.3.

Figure 6.2: A schematic illustration of structural change by thermal degradation involving crystallite and atomic migration (adapted from [7]).

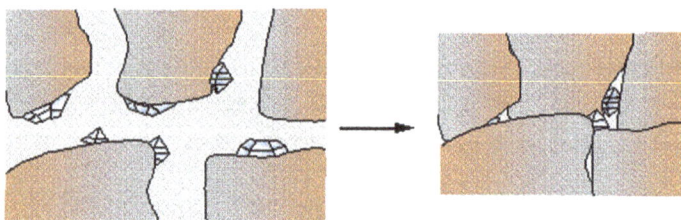

Figure 6.3: A schematic illustration of structural change by collapse of pore structure (adapted from [7]).

6.3 Deactivation by fouling

Fouling represents the physical *deposition of macroscopic material(s)* from the fluid phase onto the catalyst surface, as shown in Figure 6.4. The most common deposits are *carbon* and *coke*.

The term "coke" is a generic description of several carbonaceous deposits. Thus, deactivation by fouling is commonly called **coking**. Note that we say "most common" deposits. There are cases in which a decrease in reactor temperature could lead to condensation of liquid-like polymeric deposits on the catalyst; thus leading to a fouled surface; see [8] for an example.

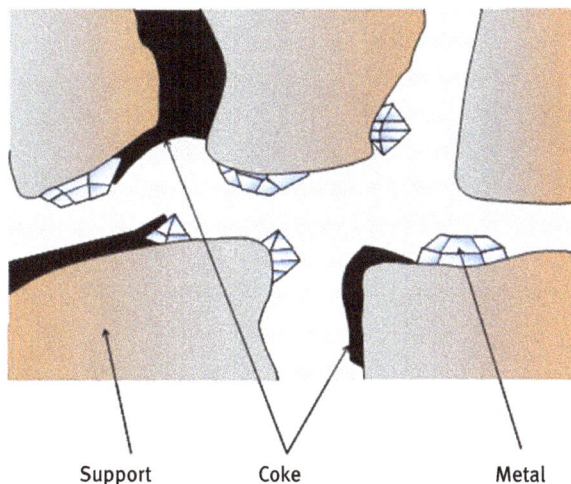

Figure 6.4: A schematic illustration of coke deposition on catalyst surface (adapted from [7]).

Just as fouling of heat exchangers causes a decline in the efficiency of the exchanger, carbon and coke deposition on the catalyst surface leads to the blockage of pores and thus to preventing the access of reactants to active sites.

The analogy with fouling in heat exchangers is meant to emphasize that coking normally manifests itself as *visible* deposits on the catalyst particle surface, as shown in Figure 6.5. Indeed, those deposits are sometimes comparable in magnitude to the weight of the catalyst particle itself, for example, 15–20 weight % [3].

Where do carbon and coke come from? Typically, carbon is a product of CO disproportionation (which is the reverse of the *Boudouard* reaction):

$$2CO_{(g)} \Leftrightarrow CO_{2(g)} + C_{(s)}$$

Still, coke could also be produced by the decomposition or condensation of heavy hydrocarbons. As shown in Figure 6.1, fouling is both a time-dependent mechanism (since reactions are involved) and a spatial mechanism (since blockage of the surface takes place).

A very important industrial process in which coking is the main cause of catalyst deactivation is the *catalytic cracking* of heavy petroleum fractions such as gas oil. An example of such a reaction is:

Figure 6.5: A photograph of fresh and spent catalyst used in the production of maleic anhydride.

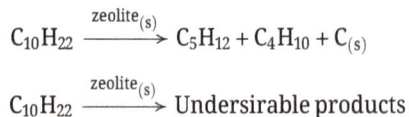

$$C_{10}H_{22} \xrightarrow{\text{zeolite}_{(s)}} C_5H_{12} + C_4H_{10} + C_{(s)}$$

$$C_{10}H_{22} \xrightarrow{\text{zeolite}_{(s)}} \text{Undersirable products}$$

As shown in Table 6.1, coking in this process is so fast that the catalyst must be *simultaneously* regenerated in another unit!

6.4 Deactivation by poisoning

A *poison* is a substance that interacts strongly with the catalyst either by adsorption or by chemical reaction. The most common cause of poisoning involves strong *chemisorption of a fluid species on the active sites*. As a result, this mechanism is classified as a chemical phenomenon in Figure 6.1. The situation is shown schematically in Figure 6.6.

You might naturally wonder where poisons come from. The answer is that poisons are normally made up of impurities that are present in *trace* amounts in the feed to the catalytic reactor. Table 6.2 lists some examples of known poisons in important industrial processes.

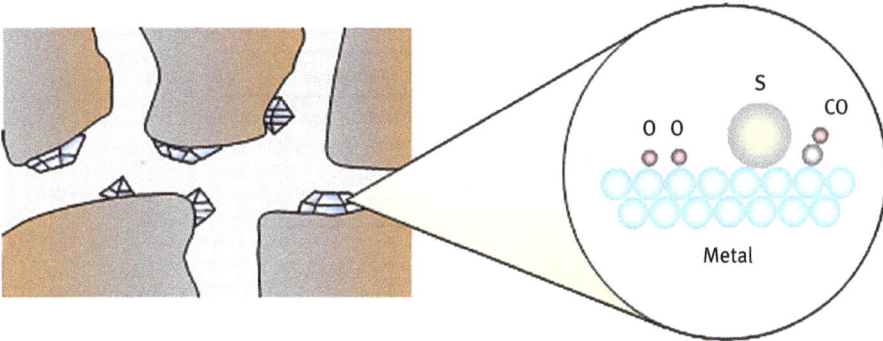

Figure 6.6: A schematic illustration of deactivation by poisoning, involving sulfur (S) (adapted from [7]).

Table 6.2: Poisons for selected catalysts in some industrial reactions (adapted from [6]).

Reaction	Catalyst	Poison(s)
Cracking	Zeolites and aluminosilicates	Organic bases and heavy metals
Hydrogenation, dehydrogenation	Ni, Pt, and Cu	Compounds of S, P, As, Zn, Hg, and Pb
Steam reforming of methane and naphtha	Ni	H_2S and As
Hydrocracking	Co and noble metals/zeolites	NH_3, S, Se, Te, and P
Oxidation of ethylene to ethylene oxide	Ag	As
Hydrotreating of petroleum residues	Co and Mo sulfides	Asphaltenes; N, Ni, and V compounds
Oxidation of CO and hydrocarbons	Pt and Pd	Pb, P, Zn, and S

6.5 Other causes of deactivation

As summarized in the introductory section, the last two causes of catalyst deactivation are **evaporation** of active catalyst component(s), and **attrition** and/or **crushing** of catalyst particles resulting in the loss of mechanical strength.

Numerous industrial catalytic processes are characterized by high reaction temperatures, and such harsh environments may lead to the evaporation of the active agent(s) and consequently a loss of the catalytic phase. Even more common is vapor transport due to the formation of *volatile compounds* involving the catalyst agent (e.g., a metal). Examples of such compounds are shown in Table 6.3.

Table 6.3: Examples of volatile compounds formed in catalytic reactions [6].

Gaseous environment	Compound type	Examples
CO and NO	Carbonyls and nitrosyls	$Ni(CO)_4$ and $Fe(CO)_5$
O_2	Oxides	RuO_3 and PbO
H_2S	Sulfides	MoS_2
Halogens	Halides	$PdBr_2$, $PtCl_4$, and PtF_6

The **mechanical failure** of catalyst particles typically occurs in the following forms:

(a) Crushing of particles, resulting from the weight of the catalyst bed in packed-bed reactors. Since industrial reactors frequently contain tons of catalyst, the catalyst particles in the bottom section of the reactor risk becoming crushed.

(b) Attrition, resulting from the abrasion of the moving catalyst particles. Industrial moving-bed or slurry reactors, and 3-phase reactive distillation units similarly contain tons of catalyst. In some cases, the solid particles are continuously recirculated between the reactor and regenerator units (see the fluidized-bed reactor in the next chapter), such that the friction between catalyst particles leads to attrition and the formation of *fines*.

6.6 Kinetics of catalyst deactivation

In the last section the different causes of deactivation were discussed qualitatively. A brief description is provided now on how catalyst deactivation is analyzed *quantitatively*. This is essential since catalytic reactor engineers are required to design and analyze the performance of equipment for catalytic reactions. The design involves the rate of reaction(s) as well as the rate of deactivation.

6.6.1 Catalyst activity

The following discussion is short and meant to demonstrate rather than provide a complete framework for the analysis of all deactivating catalyst systems.

Let us start by defining a new parameter called the **catalyst activity, a(θ)**:

$$\mathbf{a}(\theta) = \frac{\text{rate of catalytic reaction of reactant A}}{\text{rate of catalytic reaction on a fresh catalyst}} = \frac{(-r_A)}{(-r_A)_0} \qquad (6.1)$$

where θ is the **time-on-stream**, that is, the time of catalyst usage in the reactor. By using this equation and what we have learned in previous chapters, the rate of the catalytic reaction can be expressed in the following general form:

$$(-r_A) = \mathbf{a}(\theta).\eta.f(T).g(C_j) \quad j = \text{species A, B, } \ldots \tag{6.2}$$

In this equation, η is the effectiveness factor (recall Chapter 5); $f(T)$ is a temperature-dependent function that generally involves the reaction rate and adsorption constants (Chapter 4); and $g(C_j)$ is a concentration-dependent function (Chapter 4).

We are now brought face to face with the question of determining $\mathbf{a}(\theta)$. This can be done by defining the rate of deactivation [9] as follows:

$$(-r_d) = -\frac{d\mathbf{a}}{d\theta} = k_d(T).f(C_j).\mathbf{a}^d \quad j = \text{species A, B, } \ldots \tag{6.3}$$

where k_d is an Arrehenius-type deactivation constant; and d is called the order of deactivation.

Equations (6.2) and (6.3) need to be solved simultaneously so as to describe the concentration-time behavior of the system.

Example 6.1 (data adapted from [9]): We plan to run the *m*-xylene to *o*-xylene reaction in an isothermal PBR. The feed is pure *m*-xylene (A) with $F_{A0} = 5$ kmol/h, $W = 1{,}000$ kg, $P = 3$ atm, and $T = 730$ K.

The catalyst deactivates so it is planned to make 120-day test runs, then regenerate the catalyst.

Reaction rate: $(-r_A)' = 0.2$ **a** C_A^2 [mol/kg cat. h]

Deactivation rate: $-\dfrac{d\mathbf{a}}{d\theta} = 8.3125 \times 10^{-3}$ [1/day]

(a) Plot conversion of A versus time-on-stream for the whole run.
(b) What is the conversion of A at the end of the run.

Solution:
(a) Since no data is given, we will assume that $\Delta P = 0$ and $\eta = 1$.

The design equation for an isothermal PBR is given by the following equation:

$$W = F_{A0} \int_0^x \frac{dX}{(-r_{A'})} = F_{A0} \int_0^x \frac{dX}{0.2 a C_A^2}$$

In terms of conversion, $C_A = C_{A0}(1 - X)$, which can be used to solve the integral:

$$W = \left[\frac{F_{A0}}{0.2 a C_A^2}\right] \frac{X}{(1 - X)}$$

We can solve for $X = f(\theta)$ after we substitute the values of $\left[\dfrac{F_{A0}}{0.2 a C_A^2}\right]$ and W using:

$$C_{A0} = (P_{A0}/RT) = 50 \text{ mol/m}^3$$
$$\mathbf{a} = (1 - a\theta) \text{ where } a = 8.3125 \times 10^{-3}$$

Solving for X, we obtain

$$X = \frac{100(1 - a\theta)}{1 + 100(1 - a\theta)}$$

(b) The plot of X versus θ is shown below. At $\theta = 120$ days, $X = 0.2$.

coversion vs. time-on-stream

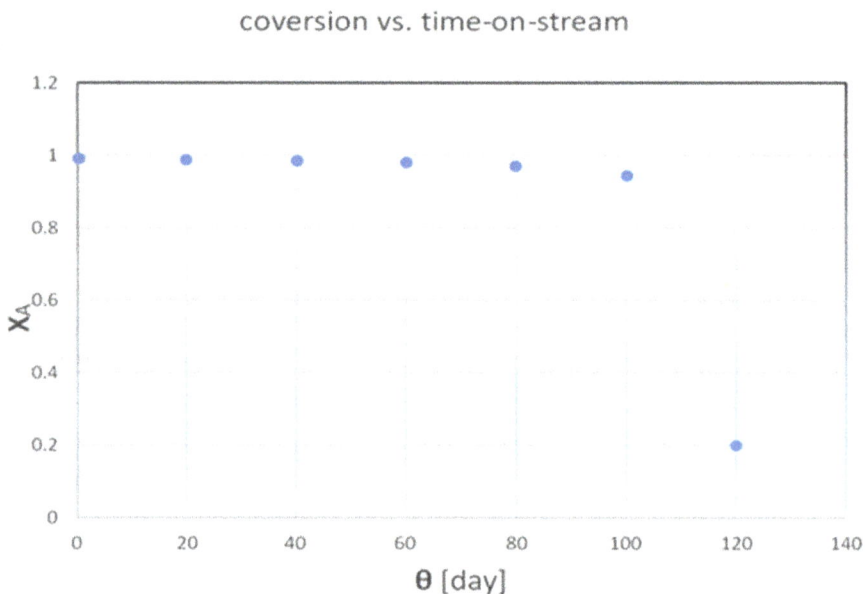

6.6.2 Kinetics of catalyst coking

The amount of coke deposited on solid catalysts is an important design parameter. This is especially important in the case of cracking reactions. Figure 6.7 shows a photograph of a fresh and spent zeolite catalyst used in the FCC process (more on FCC in Chapter 8). In this case, the amount of coke could be described by the simple *empirical* relation [10]:

$$\text{Coke} = A(\theta)^n \tag{6.4}$$

where A and n are constants that depend on the reaction, catalyst, and operating conditions [11]. According to Harriot [11], typical values of n are in the range (0.4–1.0).

Figure 6.7: A photograph of fresh and coked zeolite powder catalyst in FCC process.

6.7 Catalyst reactivation

Until the 1970s most industrial catalysts had a simple life cycle: they were either used for a single production cycle until their catalytic activity was exhausted, or for a few cycles with *in situ* regeneration between cycles, that is, within the reactor system or chemical plant itself.

More recently, the situation has changed as *off-site* regeneration (i.e., at another location) has become the preferred alternative in the chemical industry. This is due to time and safety considerations and improved recovery of the catalyst activity. In the Gulf Cooperation Council (GCC) region, few private companies provide catalyst recovery services. Those include Al-Bilad Catalyst Company Ltd. in Saudi Arabia, and Al-Manafa Group in Kuwait.[2]

The treatment of deactivated catalyst systems naturally depends on the cause of deactivation. In general, a restoration of the activity of the catalyst, or reactivation, is most desirable, which depends in turn upon reversibility of the deactivation. If the deactivation is irreversible then replacement of the catalyst is the only option. However, recovery of *expensive* catalytic species is done. Such species include the Platinum Metals Group (PMG = *Pt, Rh, Pa*, etc.).

Thermal degradation (sintering) is generally irreversible, and thus most catalyst manufacturers focus their efforts on prevention. This means that a variety of ways for slowing down this mechanism are considered in the development of catalytic materials, for example, by addition of thermal stabilizers in the manufacturing stage.

---

2 See http://www.bilad-catlyst.com.sa, and http://www.almanafa.com/catalyst_Services.html.

Fouling by coke formation is usually handled by combustive regeneration using a gas stream that contains oxygen. This procedure, carried out *in situ*, essentially restores most of the catalyst activity in reaction/regeneration cycles, as mentioned earlier.

Poisoning, on the other hand, requires knowledge of the responsible substance(s). There are thus three practical options:

(a) To treat the feed to remove the undesirable substance. An example is the hydro-desulfurization of naphtha before the naphtha reforming process in crude oil refineries.

(b) To replace the catalyst with one that has superior resistance for the poisons involved.

(c) To sacrifice the front portion of the catalyst bed in PBRs; or use a so-called *guard-bed* to trap contaminants in the inlet section of the reactor.

An example of such an arrangement is shown in Figure 6.8. The *MacroTrap® Xpore 80*, shown in Figure 6.9, is an example of a commercially available material that has a specific capacity. On the other hand, Figure 6.10 shows the *Denstone® deltaP* support medium, which has a different objective.

What do you think are the most important properties of the guard-bed media in the reactor *top* and bed-support media in the reactor *bottom*?

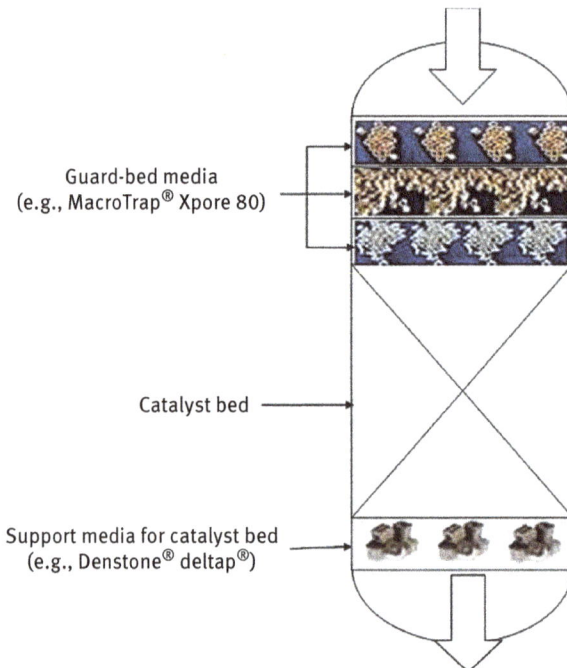

Guard-bed media (e.g., MacroTrap® Xpore 80)

Catalyst bed

Support media for catalyst bed (e.g., Denstone® deltap®)

Figure 6.8: An illustration of the guard-bed media in packed-bed reactors. (http://www.norpro.saint-gobain.com).

Figure 6.9: The *MacroTrap® Xpore 80* guard-bed medium.
Source: http://www.norpro.saint-gobain.com.

Figure 6.10: The *Denstone® deltaP* bed-support medium.
Source: http://www.norpro.saint-gobain.com.

References

[1] Doraiswamy L. K., and Sharma M. M. "Heterogeneous Reactions: Analysis, Examples, and Reactor Design. Volume 1: Gas-solid and Solid-Solid Reactions," Wiley, New York, (1984).

[2] Hagen J. "Industrial Catalysis: A Practical Approach," Wiley-VCH, Weinheim, (1999).

[3] Butt J. H., and Petersen E. E. "Activation, Deactivation, and Poisoning of Catalysts," Academic Press, San Diego, (1988).

[4] Argyle M. D., and Bartholomew C. H. "Heterogenous Catalyst Deactivation and Regeneration: A Review," Catalysis, 5, 145 (2015).

[5] Worstell J., Doll M. J., and Worstell J. H. "What's Causing Your Catalyst to Decay," Chem. Eng. Prog, 59–64 (2000).

[6] Bartholomew C. H., and Farrauto R. J. "Fundamentals of Industrial Catalytic Processes," Second Ed., Wiley, New York, (2006).

[7] Lassi U. "Deactivation Correlations of Pd/Rh Three-way Catalysts Designed for Euro IV Emission Limits: Effect of Ageing Atmosphere, Temperature and Time,"Ph.D. Thesis, Department of Process and Environmental Engineering, Oulu University, Finland, (2003).

[8] Ohlinger C., and Kraushaar-Czarnetzki B. "Improved Processing Stability in the Hydrogenation of Dimethyl Maleate to γ-butyrolactone, 1,4-butanediol and Tetrahydrofuran," Chem. Eng. Sci, 58, 1453 (2003).

[9] Levenspiel O. "Chemical Reaction Engineering," Third Ed., Wiley, New York, (1999).

[10] Wolf E. E., and Alfani F. "Catalysts Deactivation by Coking, Cat," Rev. Sci. Eng., 24(3), 329 (1982).

[11] Harriot P. "Chemical Reactor Design," Marcel Dekker, New York, (2003).

7 Industrial catalytic reactors

Beware of applied non-applicable science.
Anonymous

At this stage, we have enough knowledge about solid catalysts to *start* considering the ultimate objective of the reaction engineer, that is, to analyze, design, and operate the catalytic reactor.

This chapter defines the different types of industrial catalytic reactors and their range of applications. Focus is placed on the analysis and design of the packed-bed reactor, which is the most widely used reactor in the chemical, petroleum-refining and pollution-control industries.

7.1 Types of reactors and applications

The classic single-phase, *homogenous*, reactors that most of us first encounter in the chemical reaction engineering course include the tubular reactor[1] and the tank reactor in its closed (i.e., batch) and continuous (i.e., CSTR) forms.

Heterogeneous reactors essentially take the same basic forms, except that two phases (fluid–solid) or three phases (gas–liquid–solid) are naturally involved. However, catalytic reactors are now given different names depending on the size and shape of the catalyst particles; how the particles are placed (i.e., packed) inside the reactor; and the number of phases present in the system.

Table 7.1 lists some of the most common shapes of catalyst particles (see also Figure 1.1) and types of catalyst beds. It should be noted that the particle shapes shown represent a *few* of the many varieties possible.

The reactor applications mentioned in Table 7.1 represent the basic types of two-phase: **packed-bed** and **fluidized-bed** reactors; and three-phase catalytic reactors: **slurry** and **trickle-bed** reactors. Next, we shall describe some characteristics of the two-phase reactors; the description of three-phase reactors will be given in Section 7.3.

---

1 The homogenous tubular reactor is called the "plug-flow reactor," or PFR, in almost all chemical reaction engineering textbooks. As I tell my students, no such thing exists in the marketplace or in equipment catalogs! "Plug" flow describes an ideal situation that can be approached by controlling fluid flow and reactor dimensions.

https://doi.org/10.1515/9783111032511-007

Table 7.1: Types of catalyst particles and beds (data partially taken from [1]).

Shape	Particle dimension(s) (diameter, height, or length)	Reactor
Microspheres (powder)	$D = 0.001–0.5$ mm	– Fluid bed – Slurry
Spheres	$D = 7–100$ mm	– Packed bed – Trickle bed
Irregular granules	$D = 1–20$ mm	– Packed bed – Trickle bed
Pellets	$D = 3–15$ mm $H = 3–15$ mm	– Packed bed – Trickle bed
Extrudates	$D = 1–50$ mm $L = 3–30$ mm	– Packed bed – Trickle bed
Rings	$D_p = 20–100$ mm $H_p = 20–100$ mm	– Packed bed – Trickle bed

7.1.1 Packed-bed reactor family

The packed-bed reactor (from now on called PBR) is also commonly called a **fixed-bed reactor**. The PBR is probably the most widely applied reactor in industry. Doraiswamy and Sharma [2] list numerous examples of PBR applications in the chemical, petroleum-refining, and petrochemical industries. Indeed, almost 85% of the (two-phase) catalytic reactions, described in that reference, are conducted in PBRs and the remaining 15% are carried out in fluidized-bed reactors.

Over the years, a few PBR configurations have evolved to fit the requirements of specific reactions and operating conditions. Table 7.2 summarizes several kinds of PBR configurations, their uses, and examples of industrial applications.

Table 7.2: PBR reactor configurations (adapted from [3]).

Classification	Use	Typical applications	Description
Single bed	Moderately exothermic or endothermic, non-equilibrium limited, reactions	Mild hydrogenation	Figure 7.1
Multi-tubular	Highly endothermic or exothermic reactions requiring close temperature control to ensure high selectivity	C_2H_4 oxidation to C_2H_4O; CH_3OH oxidation to formaldehyde	Figure 7.2
Beds in series with inter-stage cooling/heating	Equilibrium-limited reactions	Catalytic reforming; NH_3 synthesis	Figure 7.3
Radial flow	Where low pressure drop is required, and when change in moles is large	Styrene from ethyl-benzene	Figure 7.4
Direct-fired	Highly endothermic; high- temperature reactions	Steam reforming of natural gas	See [4]

Figures 7.1–7.4 schematically illustrate the physical placement of the catalyst bed(s) within the reactor.

The family of PBRs therefore includes:
(a) reactor with a single bed;
(b) multi-tubular reactor with shell-and-tube arrangement;
(c) reactor involving beds in series with inter-stage cooling or heating
(d) radial flow reactor; and
(e) direct-fired reactor.

Of course, the development of the PBR has continued and new technologies have evolved beyond those given in Table 7.2. An example of such a new technology is the *Linde reactor* (see Isothermal reactor | Linde Engineering (linde-engineering.com)).

Feed

Inert beads

Catalyst bed

Inert beads

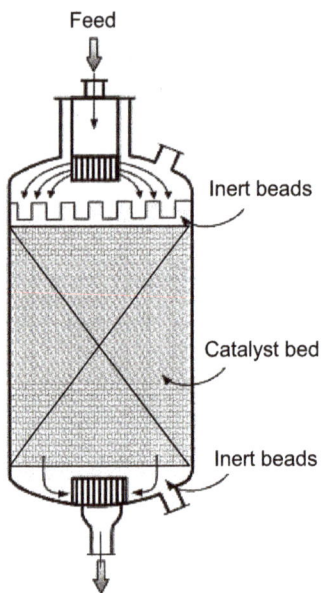

Figure 7.1: A schematic illustration of the catalytic packed-bed reactor.

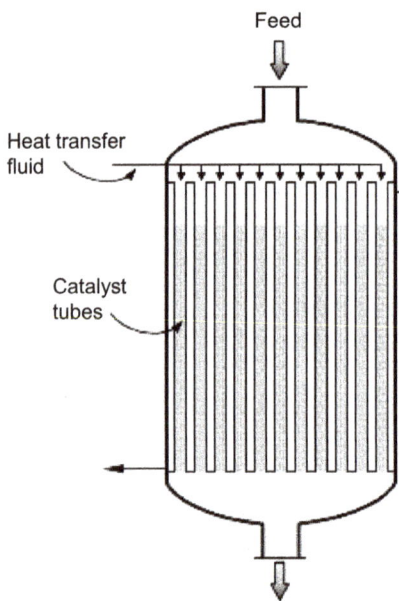

Feed

Heat transfer fluid

Catalyst tubes

Figure 7.2: A schematic illustration of the multi-tubular catalytic packed-bed reactor.

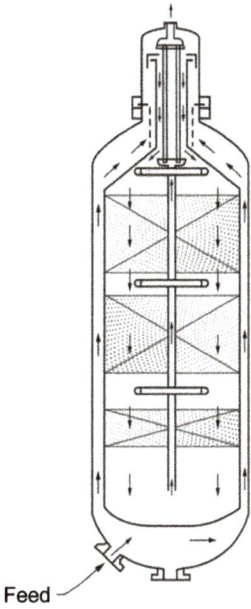

Figure 7.3: A schematic illustration of the packed-bed reactor with three beds in series with inter-stage heat transfer.

Feed

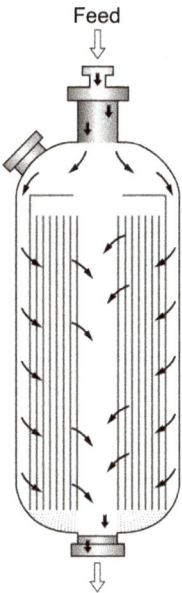

Figure 7.4: A schematic illustration of the radial-flow packed-bed reactor (adapted from [3]).

Feed

7.1.2 Fluidized-bed reactor family

As listed in Table 7.1, the fluidized-bed reactor (also commonly called *fluid-bed reactor*) hosts extremely small catalyst particles (almost powder-like micro-spheres). The term "fluidized" refers to the state of motion of the solid particles, which behave like a fluid, in such reactors. Most industrial applications of this reactor involve gas-solid systems, and such contacting has been called "aggregative" fluidization.[2]

There are several possibilities for gas-solid contacting, arising from low to very high feed superficial gas velocities (U_{go}), that lead to various types of fluidized-bed reactors. These possibilities are schematically depicted in Figure 7.5.

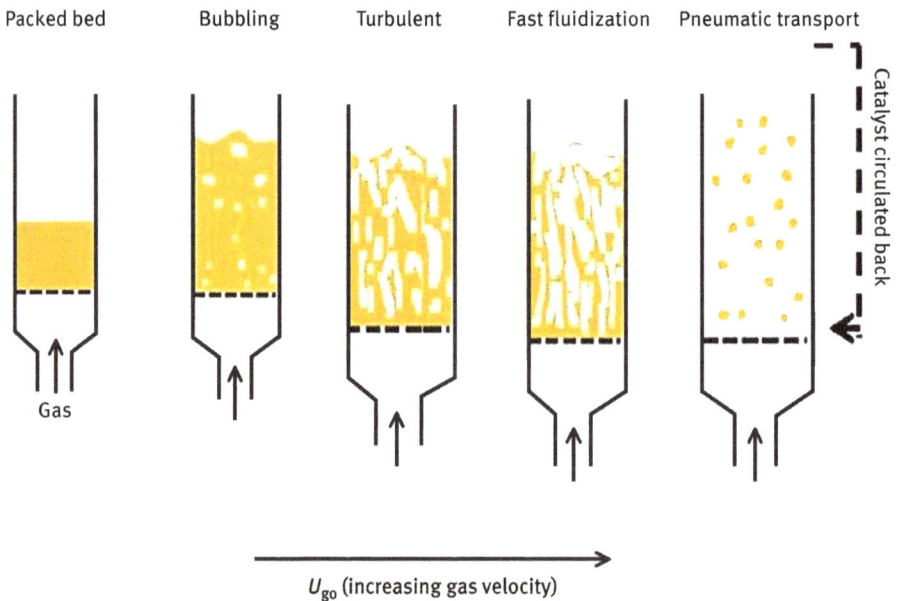

Figure 7.5: Various types of contacting in gas-solid fluidization.

The family of fluidized-bed reactors therefore includes [5]: (a) bubbling fluidized bed (BFB); (b) turbulent fluidized bed (TFB); (c) fast fluidized bed (FFB); and (d) pneumatic conveying bed (PCB) which is also called the *transport-line reactor* (TLR).

In industrial applications, the BFB and TLR probably make up the most important modes of operation. In the BFB reactor, the solid particles are kept in a well-mixed **suspended** state to ensure that good heat transfer is accomplished in highly exothermic reactions.

---

2 Liquid-solid contacting is called "particulate" fluidization.

In the TLR, the particles are **entrained** with the gas stream, that is, the solid particles leave the reactor. This is done to *regenerate* the rapidly deactivated catalyst in another unit (recall Chapter 6). As shown in Figure 7.5, the regenerated solids are returned to the reactor, thus giving rise to the so-called **circulating fluidized bed** (CFB) mode of operation.

A very important industrial application of the TLR[3] and CFB technology is the fluid catalytic cracking (FCC) process in petroleum refineries. FCC is used to produce hydrocarbon fractions in the gasoline and diesel boiling range from "bottom cuts" (mainly gas oil) produced in atmospheric or vacuum crude oil distillation towers.

7.2 Design of isothermal packed-bed reactor

When considering the subject of isothermal, catalytic, reactor design, we are usually interested in answering questions such as:
(a) Given data on the reaction kinetics, catalyst particle, and product specifications, how much catalyst is needed, and what is the required reactor volume?
(b) Given data on the reaction kinetics, catalyst particle, and a fixed amount of catalyst, what is the expected conversion of the limiting reactant and what is the product distribution (i.e., yield and selectivity in multiple reaction systems)?
(c) When attempting to answer (a) and (b), how significant are internal and external mass transfer effects in the reaction system under consideration?

This section focuses attention on questions (a) and (b) by developing and applying the *performance equation*[4] for an isothermal, steady-state, PBR. As for question (c), we fall back on the material presented in Chapter 5.

Let us now consider the catalytic reaction A + B → Products. In the case of solid catalytic reactions, the amount of catalyst is what is important for product formation. Therefore, it is more suitable to define the bulk rate of the catalytic reaction based on catalyst mass (W) as follows:[5]

$$(-r'_A)_b = \frac{\text{moles of A consumed}}{\text{time} \cdot \text{mass of catalyst}} \tag{7.1}$$

The bulk rate of the reaction per mass of catalyst can be related to *reaction rate per volume of reactor*, $(-r_A)$, using the **bulk bed density**, ρ_B (defined as mass of catalyst/ volume of reactor), so that we obtain the useful relationship:

---

3 The transport-line reactor is called the "riser" in the refinery FCC process.
4 Recall that the performance equations for *homogeneous* reactors are alternatively called the "design equations" in standard textbooks; see, for example [6], and [7].
5 This corresponds to the *bulk* rate of reaction $(-r_A)_b$ introduced in Chapter 5.

$$(-r_A) = \rho_B(-r'_A)_b \tag{7.2}$$

The development of the performance equation for an isothermal PBR, such as that shown in Figure 7.1 or Figure 7.2, is based on three key assumptions:

(1) *The fluid phase is in plug flow*; that is, the velocity profile is flat across the bed.
(2) *A uniform distribution of flow over the bed cross-section* (see the top part of the reactor in Figure 7.1 for an example of how this could be accomplished in practice).
(3) The heterogeneous nature of the system may cause appreciable concentration differences between the bulk fluid and internal and/or external surface of the catalyst particles. These differences are accounted for using the *effectiveness factor* concept.

Referring to Figure 7.6, we can write a steady-state mole balance for component A, over the differential element ΔW, as follows:

$$F_A|_W - F_A|_{W+\Delta W} = \eta(-r'_A)_b.\Delta W \tag{7.3}$$

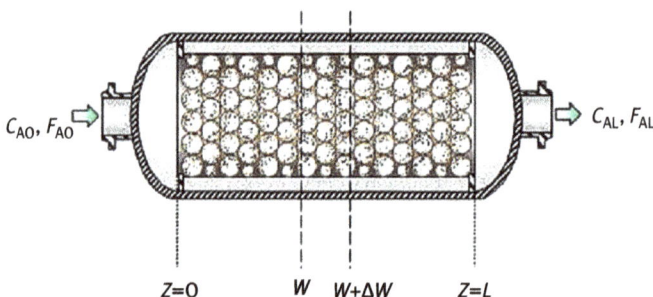

Figure 7.6: The mole balance in a differential "slice" of a packed-bed reactor.

where F_A is the molar flow rate of component A, and η is the effectiveness factor. Taking the limit of $\Delta W \to 0$ after rearranging eq. (7.1), we obtain the **differential form** of the performance equation:

$$-\frac{dF_A}{dW} = \eta(-r'_A)_b \tag{7.4}$$

Further assuming that
(4) *Catalyst deactivation is insignificant*, and
(5) *Pressure drop across the catalyst bed is negligible*,

we can integrate eq. (7.3) to obtain the **integral form** of the performance equation:

$$W = - \int_{F_{A0}}^{F_A} \frac{dF_A}{\eta(-r'_A)_b} \tag{7.5}$$

We can also express the performance equations in terms of the conversion of reactant A, defined as $X_A = (F_{A0} - F_A)/F_{A0}$. We therefore obtain:

$$\frac{dX_A}{dW} = \frac{\eta(-r'_A)_b}{F_{AO}} \tag{7.6}$$

$$W = F_{AO} \int_0^{X_A} \frac{dX_A}{\eta(-r'_A)_b} \tag{7.7}$$

7.2.1 Cautionary remarks

At this stage, it is important to reflect on the last two assumptions (4) and (5) in light of what we have learned about catalysts and our engineering background.

While catalyst deactivation is unavoidable, as was demonstrated in Chapter 6, there are several applications in which it is relatively slow (e.g., 2–3 years). In those cases, catalyst activity can be taken as essentially constant, and assumption (4) is, thus, not unreasonable.

Assumption (5), however, is somewhat hard to justify, except in very few situations. While the pressure drop in (empty) homogenous reactors is mostly unimportant, our background in fluid transport teaches us that the pressure drop in packed beds is generally significant.

Accordingly, PBR design calculations should consider the variation of pressure along the reactor, as described by the **Ergun equation.** The design problem then becomes more complicated, and in most situations, we need to combine eq. (7.4) or (7.6) with the Ergun equation and solve a set of differential equations numerically. Details of such an analysis and examples are clearly presented in mainstream reaction engineering textbooks (see, e.g. [7, 8],) and are not therefore presented here. In any case, we shall look at non-isobaric PBRs among the computational case studies in Chapter 9.

A third point of concern is related to catalytic reaction rates. When searching for catalytic reaction rates, one needs to be careful as to the *basis* of the data reported. *Intrinsic* reaction rate data are sometimes presented on the basis of *mass* of catalyst particle, or the *volume* of particles. Other times, the rate is reported per *volume of reactor*. As a result, it is necessary to distinguish between the **density of the particle**, ρ_p, and the **bulk density of the bed**, ρ_B, used in eq. (7.2). The two densities are related as follows:

$$\rho_B = (1 - \epsilon_B)\rho_p \tag{7.8}$$

where ϵ_B is the **bed porosity** (or **bed void fraction**), defined as the volume of voids/volume of reactor.[6] Thus, the fraction of solids in the reactor, that is, the volume of catalyst/volume of reactor = $(1 - \epsilon_B)$.

Example 7.1: The reaction $A + B \rightarrow R + S$ takes place isothermally in an experimental PBR. The gaseous feed enters the reactor at a rate of 10 m³/h, where $C_{A0} = 0.1$ mol/m³ and $C_{B0} = 10$ mol/m³. Calculate the mass of catalyst required for 91% conversion of A, given that the reaction is second order and $(-r'_A) = 0.6\ C_A C_B$, in units (mol/kg.h). Mass transfer effects can be ignored.

Solution: Equation (7.7) can be used to calculate the amount of catalyst required:

$$W = F_{A0} \int_0^{X_A} \frac{dX_A}{\eta(-r'_A)} = F_{A0} \int_0^{0.91} \frac{dX_A}{(1)kC_A C_B}$$

Since there is no change in moles upon reaction, we note that it is possible to use the simple stoichiometric relationship between C_B and X_A. Noting that the operation is isothermal, the above integral simplifies to:

$$W = \frac{F_{A0}}{k(C_{A0})^2} \int_0^{0.91} \frac{dX_A}{(1 - X_A)(100 - X_A)}$$

As $X_A \in [0, 1]$, we can approximate the term $(100 - X_A) \approx 100$, and integrate as follows:

$$W = \frac{F_{A0}}{k(C_{A0})^2} \int_0^{0.91} \frac{dX_A}{100(1 - X_A)} = 4 \text{ kg}$$

Example 7.2 (adapted from [9]): The first-order irreversible reaction $A \rightarrow B$ is conducted isothermally in a PBR. The reaction takes place at 450 K using 0.3 cm radius spherical particles whose density is $\rho_p = 0.85$ g/cm³. The reaction rate constant was reported to be $k = 2.61$ s⁻¹ at 450 K. The effective diffusivity of A in the catalyst particles is $D_e = 0.007$ cm²/s. Calculate the catalyst mass and reactor volume needed to obtain 97% conversion. The feed is pure A at 1.5 atm and its flow rate is 12 mol/s. The bed density has been estimated to be $\rho_B = 0.6$ g/cm³. Assume that external mass transfer limitations are negligible.

Solution: Equation (7.5) can be used to find the required amount of catalyst, but first we need to calculate the effectiveness factor. Since there is no external mass transfer resistance, we can use the expressions given in Table 5.1, Chapter 5, for the Thiele modulus and effectiveness factor:

$$\phi = \sqrt{\frac{kR^2}{D_e}} = 5.8$$

---

6 Do not confuse ϵ_B with particle porosity that was defined in Chapter 3.

$$\eta = \frac{3}{\phi}\left(\frac{1}{\tanh\phi} - \frac{1}{\phi}\right) = 0.428$$

Before applying eq. (7.5), we note that the units of k, *reported in this example*, are not consistent with the units of k according to the definition of $(-r_A')$. Therefore, it is important to reconcile the units and divide k by ρ_p. Equation (7.5) then yields the following value of W, where the ideal gas equation of state can be used to calculate the initial concentration of A:

$$W = F_{A0}\int_0^{X_A}\frac{dX_A}{\eta(-r_A')} = \frac{F_{A0}}{\eta\left(\frac{k}{\rho_p}\right)C_{A0}}\int_0^{0.97}\frac{dX_A}{(1-X_A)} = \frac{-F_{A0}}{\eta\left(\frac{k}{\rho_p}\right)C_{A0}}\ln(1-X_A)\big|_0^{0.97} = 789\ \text{kg}$$

The required reactor volume can be computed using the definition of the bed density:

$$V = \frac{W}{\rho_B} = 1.32\ \text{m}^3$$

7.3 Design of isothermal transport-line reactor

The fluidized-bed reactor is second in importance to the PBR. Its design, in its various operating modes, is beyond the scope of this book. However, a special operating mode of the fluidized-bed reactor, namely, *pneumatic transport* (see Figure 7.5) is somewhat amenable to relatively simple mathematical analysis, where exact solutions can be obtained in limiting cases.

The reactor in that mode of operation has been given different names, including pneumatic-transport reactor, recirculating solid riser (or simply the *riser*), transfer-line reactor, and **TLR**, which we shall henceforth adopt.

The importance of TLR is a result of the fact that it is the main reactor in the FCC process in the refining industry, as we shall see in Chapter 8. By the way, there are more than 400 units of the TRL in the world.

In the derivation of the design equation for the isothermal TLR, we invoke the following assumptions:
(1) The fluid and solid phases are in plug flow; that is, the velocity profile is flat across the reactor.
(2) The TRL reactor typically contains powder-sized solid catalysts (the particle diameter is in the range of 10–100 μm) such that diffusional effects are assumed negligible, that is, $\eta \rightarrow 1$.
(3) According to Chapter 5, a reasonable catalyst *decay function* is given by the following equation, when rapid coking occurs (as in the FCC process):

$$a = \frac{k}{k_0} = \exp(-\alpha\theta) \tag{7.9}$$

where α is the decay constant and θ is time-on-stream (or catalyst residence time).

Referring to Figure 7.7, we can write a steady-state mole balance for reactant A, over the differential element ΔZ (details of the derivation are given in the textbook website at www.degruyter.com). The mole balance leads to the relatively simple design equation:

$$\frac{dX_A}{d\theta} = \left[\frac{(1-\epsilon)U_s}{\epsilon U_g}\right]\frac{(-r_A)'}{C_{A0}} \tag{7.10}$$

where U_s is solid catalyst particle velocity, U_g is gas velocity, and ϵ is reactor void fraction.

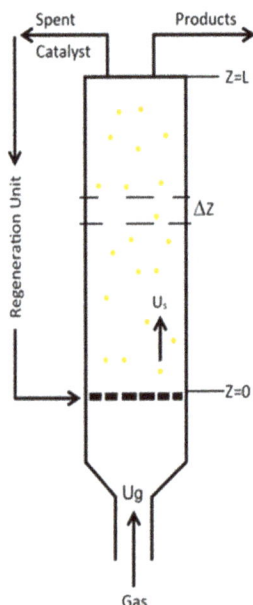

Figure 7.7: The mole balance in a differential slice of the transport-line reactor.

7.4 Three-phase catalytic reactors

A three-phase catalytic reactor is a vessel in which gas and liquid phases are in contact with the presence of a solid catalyst. Most industrial cases involve a reaction between a dissolved gas and a liquid phase at the solid surface. In other cases, the liquid is an inert medium and the reaction takes place between dissolved gases [10]. In all cases, however, the *absorption* of the gas (or gases) into the liquid phase is a prerequisite for the reaction.

Figure 7.8 shows the steps involved in gas-liquid-solid reactions. Compared to what we have seen in Chapter 1 on the number of steps in catalytic two-phase reactions, we now have *two* more diffusional steps.

It is useful to compare Figure 7.8 to Figure 1.8 to identify the numbers of those two additional mass-transfer steps in this figure.

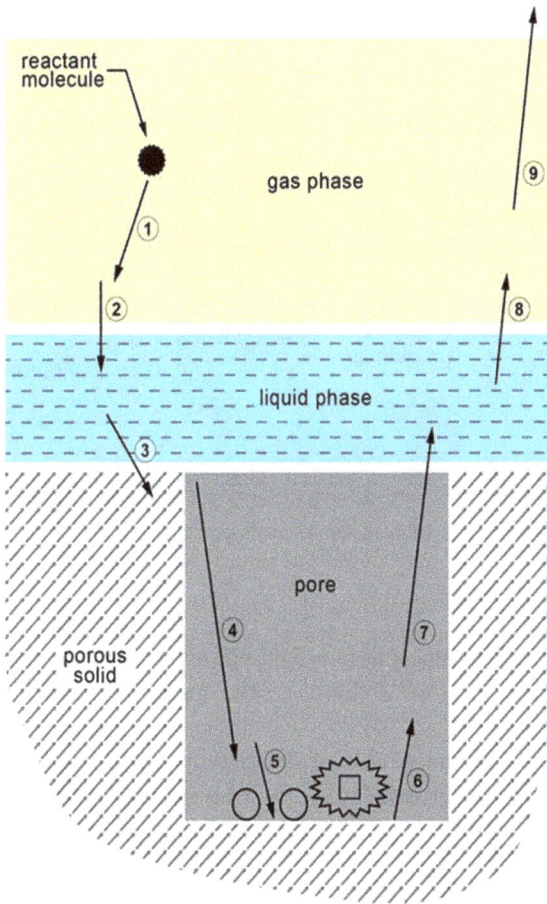

Figure 7.8: A schematic of the steps involved in a three-phase reaction in porous solid catalysts.

Industrial three-phase reactors have been classified into two main categories:

a. Slurry reactors (SRs)

In this type of reactor, (tiny) catalyst particles are suspended in the liquid and are in continuous motion. Figure 7.9 displays two types of slurry reactors. Note that the reactors are given names that correspond to the mechanism by which the catalyst particles are suspended.

An important application of the SR is the hydrogenation of vegetable oils to produce solid margarine (a vegetarian form of butter). In this reaction, *un*saturated fats in liquid vegetable oil are hydrogenated by bubbling hydrogen gas through the oil in the presence of a solid catalyst (Raney-Ni). The produced saturated fat, called margarine, has a higher melting point compared to the original oil.[7]

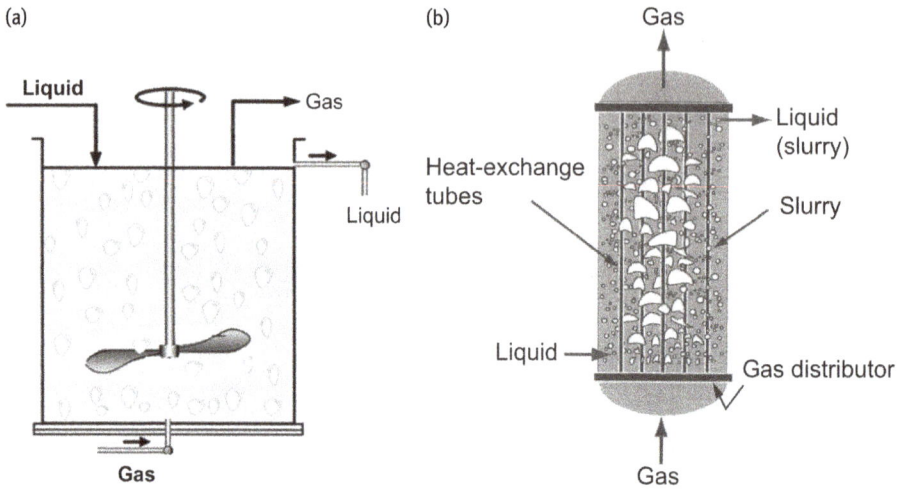

Figure 7.9: (a) Mechanically agitated slurry reactor. (b) Bubble column slurry reactor.

b. Trickle-bed reactors (TBRs)

Here, the situation is like that in PBRs except that two fluid phases now move over the stationary bed of catalyst particles. Figure 7.10 illustrates the operation in trickle-bed reactors.

TBRs are frequently used when it is undesirable to vaporize the liquid reactant. In oil refineries, for example, the TBR is applied for the hydro-treatment of "heavy cuts" from crude oil distillation towers, since vaporizing these fractions would lead to unwanted side products.

Many more examples of industrial applications of three-phase reactors can be found in Doraiswamy and Sharma [11].

---

7 This process has lost its popularity because of health concerns (you can read about those health concerns at www.medicalnewstoday.com/articles/hydrogentated-oil).

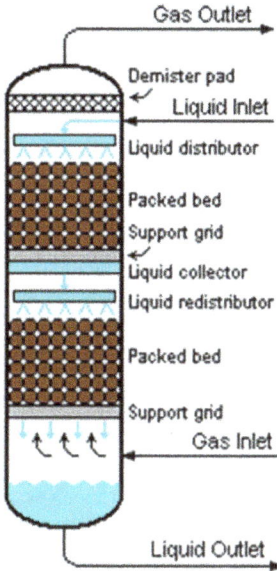

Figure 7.10: A three-phase counter-current trickle-bed reactor. (Source: Packed bed – Citizendium).

References

[1]   Turaga U., Engelbert D. R., Beever W. H., Osbourne J. T., Wagner B., Allen R., and Braden J. "Succeed at Catalyst Scale-up," Chem. Engg. Prog., 29–33 (2006).
[2]   Doraiswamy L. K., and Sharma M. M. "Heterogeneous Reactions: Analysis, Examples and Reactor Design. Volume 1: Gas-Solid and Solid-Solid Reactions," Wiley, New York, (1984).
[3]   Rase H. F. "Fixed-Bed Reactor Design and Diagnostics – Gas-Phase Reactions," Butterworths, Boston, (1990).
[4]   Froment G. F., and Bischoff K. B. "Chemical Reactor Analysis and Design," Second Ed., Wiley, New York, (1990).
[5]   Kunni D., and Levenspiel O. "Fluidization Engineering," Second Ed., Butterworth-Heinemann, Boston, (1991).
[6]   Levenspiel O. "Chemical Reaction Engineering," Third Ed., Wiley, New York, (1999).
[7]   Fogler H. S. "Elements of Chemical Reaction Engineering," Fifth Ed., Prentice Hall, Boston, MA, (2016).
[8]   Hill C. G. Jr., and Root T. W. "Introduction to Chemical Engineering Kinetics & Reactor Design," Second Ed., Wily, New Haven, NJ, (2014).
[9]   Rawlings J. B., and Ekerdt J. G. "Chemical Reactor Analysis and Design Fundamentals," Nob Hill Publishing, Madison,WI, (2002).
[10] Ramachandaran P. A., and Chaudhari R.V. "Three-Phase Catalytic Reactors," Gordon and Breach, New York, (1983).
[11] Doraiswamy L. K., and Sharma M. M. "Heterogeneous Reactions: Analysis, Examples and Reactor Design," Volume 2: Fluid-Fluid-Solid Reactions," Wiley, New York, (1984).

8 Major industrial applications

> The most important reactor by far in 20th century technology is the fluidized catalytic cracker.
> Lanny D. Schmidt

Solid catalysts are of central importance in the global industries of petroleum refining, petrochemicals, and transportation. Catalysts are also effectively used for air pollution control, especially pollution from mobile sources including mainly cars.

Indeed, a good percentage of the *Nobel Prizes in Chemistry* were related to industrial applications of catalysts. The list of laureates includes *Ostwald* (HNO_3 manufacture, 1909); *Haber* (NH_3 synthesis, 1918); *Ziegler* and *Natta* (polyethylene and polypropylene synthesis, 1963); and *Heck, Negishi*, and *Suzuki* (Pd-catalyzed cross coupling reactions, 2010).

In 2004, the worldwide solid catalyst market was estimated at $\$10.5 \times 10^9$ [1]. In 2019, that number has grown to $\$33.9 \times 10^9$ [2].

Early in the book, a classification of the major applications of solid catalysts was given in Figure 1.2. Those applications include in decreasing percentage of applications: environmental, chemical synthesis, polymers/petrochemicals, and petroleum refining.

The objective of this chapter is to present examples of some of the major industrial applications of solid catalysis. The discussion will be broad and limited to applications in petroleum refining, production of petrochemicals, and air-pollution control. These applications represent by far the most common applications of solid catalysts in the GCC countries and probably most industrial countries.

8.1 Applications in petroleum refining

Petroleum is presently the major source to produce *commodity chemicals* (chemicals produced in large quantities, e.g., NH_3 and C_2H_4O) and *transportation fuels*.

As we know, petroleum refining consists of the conversion of crude oil to a variety of products ranging from gasoline for cars, ethylene for the production of plastics, and asphalt for use on roads.

There are 731 refineries in the world with a capacity greater than 91.8×10^6 barrel/day in 2021 [2]. In Saudi Arabia, the largest petroleum producer in the world, there are nine refineries with a combined throughput capacity of around 2.83×10^6 barrel/day [3].

A flowchart of a typical refinery is presented in Figure 8.1. As is well known, the first process in the refinery is the separation of the crude oil into light fractions, intermediates, and heavier fractions of a wide range of molecular weights. The separation is achieved by atmospheric and vacuum distillation, as the higher molecular-weight species have higher boiling points. Table 8.1 lists the major products of crude oil distillation.

https://doi.org/10.1515/9783111032511-008

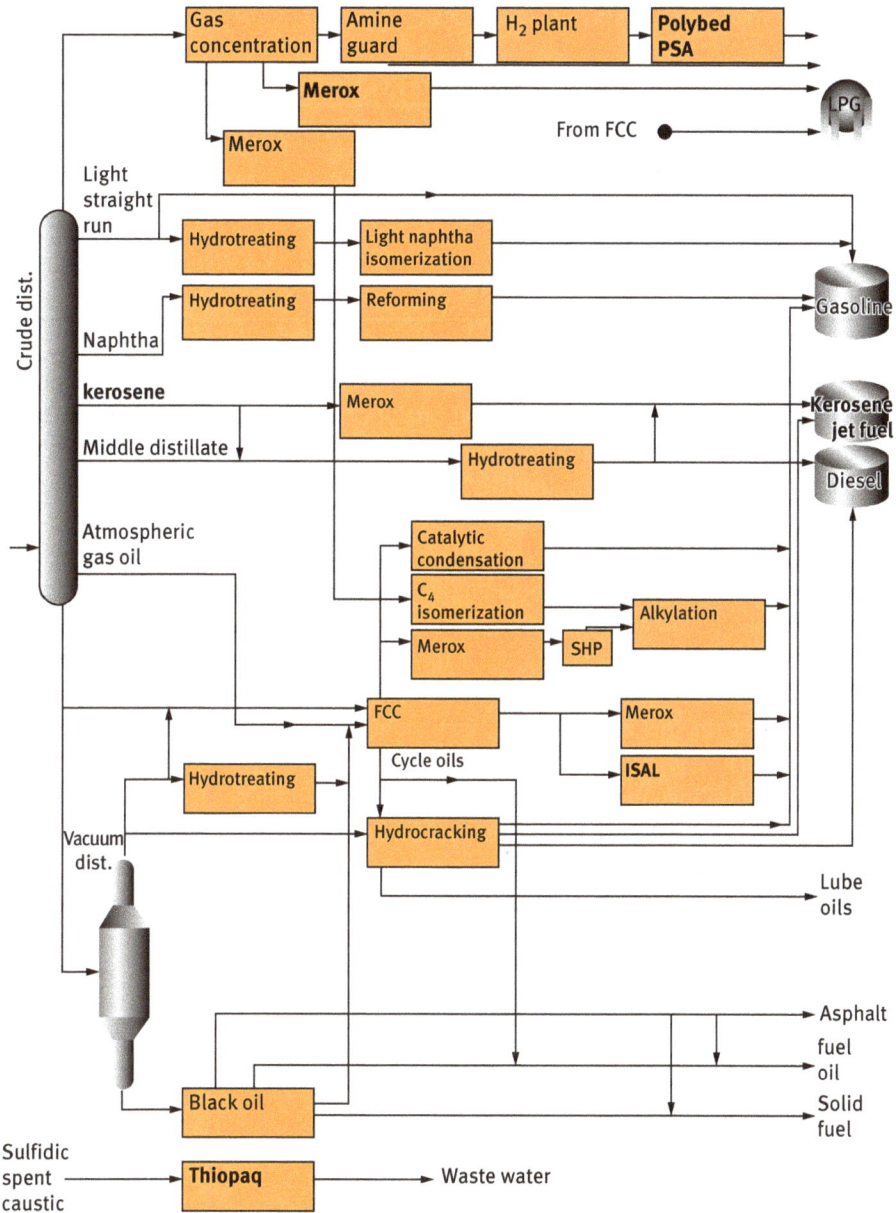

Figure 8.1: A flowchart of a typical refinery (www.uop.com).

Table 8.1: The major constituents of crude oil distillation.

Fraction	Distillation temperature (°C)	Carbon number
Butane and lighter gases	<30	C_1–C_4
Gasoline blending species	30–85	C_5–C_6
Naphtha	85–175	C_6–C_7
Kerosene; jet fuel	175–230	C_{12}–C_{18} and aromatics
Distillate (diesel; heating oil)	230–345	C_8–C_{24}
Heavy gas oil	345–565	Long chains attached to cyclic structures
Asphalt (solid)	>565	Polycyclic structures

Hydrotreating of naphtha and other distillation tower "cuts," such as the naphtha hydrodesulfurization process[1] are essential catalytic treatment processes in all refineries. In this process, sulfur compounds (and other undesirable poisoning compounds) are reduced to acceptable limits, in order to avoid a rapid catalyst *deactivation* in subsequent refinery operations.

The most useful molecules for chemical synthesis and for transportation fuels are produced by catalytic processes, including *hydrocracking, reforming,* and *cracking* of the higher chain-length alkenes. This leads to the production of unsaturated molecules closer to C_8, and this is important due to the large worldwide demand for gasoline. Some of the molecules, particularly cycloalkanes (also called cyclic paraffins), are reformed. Once again, this is to produce functionality by hydrogen loss and to open the rings [4].

This way gasoline(s) with a high-octane *rating* is/are produced. The octane rating (or octane number) is a measure of the performance of gasoline fuels used in spark-ignition internal combustion engines.[2]

Examples of some of the solid catalysts involved in these processes are listed in Table 8.2.

Table 8.2: Examples of catalysts used in refinery processes [5].

Process	Catalyst
Hydrodesulfurization	Co/Mo sulfides/γ-Al_2O_3
Reforming	Pt/η-Al_2O_3
Hydrocracking	Pd/zeolite
Fluid catalytic cracking	Zeolites

---

1 Recall the HDS catalyst in Chapter 1.
2 The higher the number, the better the fuel burns within the vehicle engine.

In previous chapters, we mentioned the **fluid catalytic cracking (FCC)** process. Schmidt [6] describes this refinery process as the *"most important reactor by far in 20th century technology!"*. There are around 400 FCC units in the world. The reason is that FCC continues to be the most economic process to convert vacuum gas oil (VGO) and other heavy residues (the so-called "bottom of the barrel") to gasoline and C_2–C_4 olefins.

Let us thus briefly describe what is involved in FCC, for which a typical flowchart is shown in Figure 8.2. In this process, there are *two* catalytic reactors: a transport-line reactor (called the *riser* in the refining industry) and a fluidized-bed reactor (called the *regenerator*).

Vaporized "heavy" gas oil is fed to the riser along with the solid catalyst powder. Cracking is accomplished in the riser where the catalyst is rapidly deactivated in a few seconds due to coking. A photograph of the FCC zeolite catalyst in its fresh and deactivated (spent) forms is given in Figure 6.6.

The deactivated catalyst is then separated from the products in a cyclone and sent to the regenerator where air is used to oxidize and remove most of the coke. Subsequently, the regenerated catalyst is sent back to the cracking reactor unit.

The cracking reactions are endothermic, whereas the combustion reactions in the regenerator are exothermic. As Schmidt [6] further says the riser-regenerator combination *"solves the heat-management and coking problems simultaneously."*

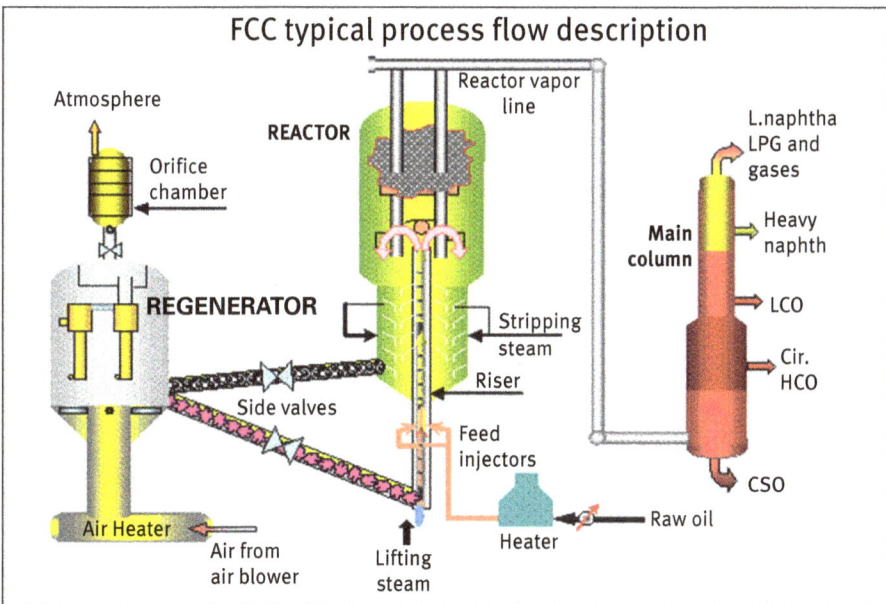

Figure 8.2: Typical process flow diagram of the FCC process.

8.2 Applications in the petrochemicals industry

The major feedstocks in the petrochemicals industry are *natural gas* (mainly CH_4 and smaller amounts of C_2H_6, C_3H_8, and higher alkanes) and *short-chain alkenes* produced in refineries.

Natural gas is abundant in the GCC countries. Consequently, giant industrial corporations (e.g., SABIC; ADNOC; Qatar Energy) and massive industries have been founded over the last 4–5 decades.

One such industry involves **steam reforming of natural gas** to produce *synthesis gas* (or *syngas*), which is mainly used for the manufacture of methanol and ammonia. These two major chemicals are in turn employed to produce other important chemicals, as described below.

In steam reforming, methane is heated to a high temperature as it passes through a packed-bed reactor (the typical catalyst is Ni/Al_2O_3). The reaction involves the oxidation of methane and the production of hydrogen:

$$CH_4 + H_2O \rightarrow CO + 3\ H_2$$

The reaction is difficult to complete in one step and usually requires a secondary reactor to oxidize the remaining methane:

$$CH_4 + O_2 \rightarrow CO + 2\ H_2$$

The *water gas shift reaction* occurs during this process to form CO_2 as the major product, utilizing another packed-bed reactor (Fe_2O_3 and Cr_2O_3 are employed as catalysts), in addition to more hydrogen:

$$CO + H_2O \rightarrow CO_2 + H_2$$

The manufacture of *methanol* in steam reforming plants also occurs catalytically ($CuO + ZnO/Al_2O_3$ as the catalyst) according to the reaction:

$$CO + 2\ H_2 \rightarrow CH_3OH$$

In a subsequent step, methanol is converted to many products, most of which are produced in catalytic processes, as shown in Figure 8.3.

On the other hand, petroleum refineries represent a major source of *short-chain alkenes*,[3] mainly olefins such as C_2H_4, C_3H_6, and C_4H_8. As shown in Figure 8.4, most major petrochemicals are produced from these basic sources, often using a variety of catalytic processes.

---

3 Short-chain alkenes include C_nH_{2n}, where n is a small integer. These include ethene, propene, and butene. But we will occasionally follow the more common names in the refinery and petrochemical industries: ethylene, propylene, etc.

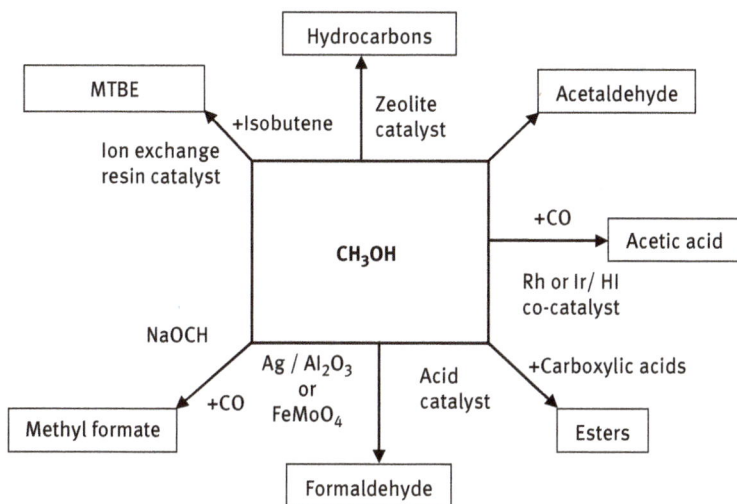

Figure 8.3: Catalytic processes in conversion of methanol to other products.

Ethylene and propylene are thus the basis of a great number of our modern materials, namely polyolefins, such as **polyethylene** and **polypropylene**. These represent what we commonly call **plastics**.

As also shown in Figure 8.4, ethylene is converted to *ethylene oxide* using an Ag/Al_2O_3 catalyst. Ethylene oxide is in turn employed on a large scale to produce many other chemicals for modern applications. For example, ethylene oxide produces ethylene glycol that is used in "antifreeze" solutions in car radiators. Another example is ethanol-amines, which are important liquid solvents for "gas sweetening" operations in natural gas plants in the oil industry.

8.3 Applications in environmental protection

Air and water pollution originate from many sources such as fossil fuel consumption and CFC[4] emissions. Consequences on human beings and the environment include: breathing problems and other illnesses, "acid rain," "global warming," and the "ozone hole" phenomenon.

The largest fossil fuel consumers, and thus the largest sources of air pollution by far, are *motor vehicles*. For this reason, laws were enacted in industrialized countries some 40 years ago with the aim of reducing **automobile emissions**. These laws (e.g.,

---

4 Chlorofluorocarbons (CFCs) are paraffin hydrocarbons that contain only carbon (C), hydrogen (H), chlorine (Cl), and fluorine (F); produced as volatile derivatives of methane, ethane, and propane. CFCs are commonly known by the brand name "Freon."

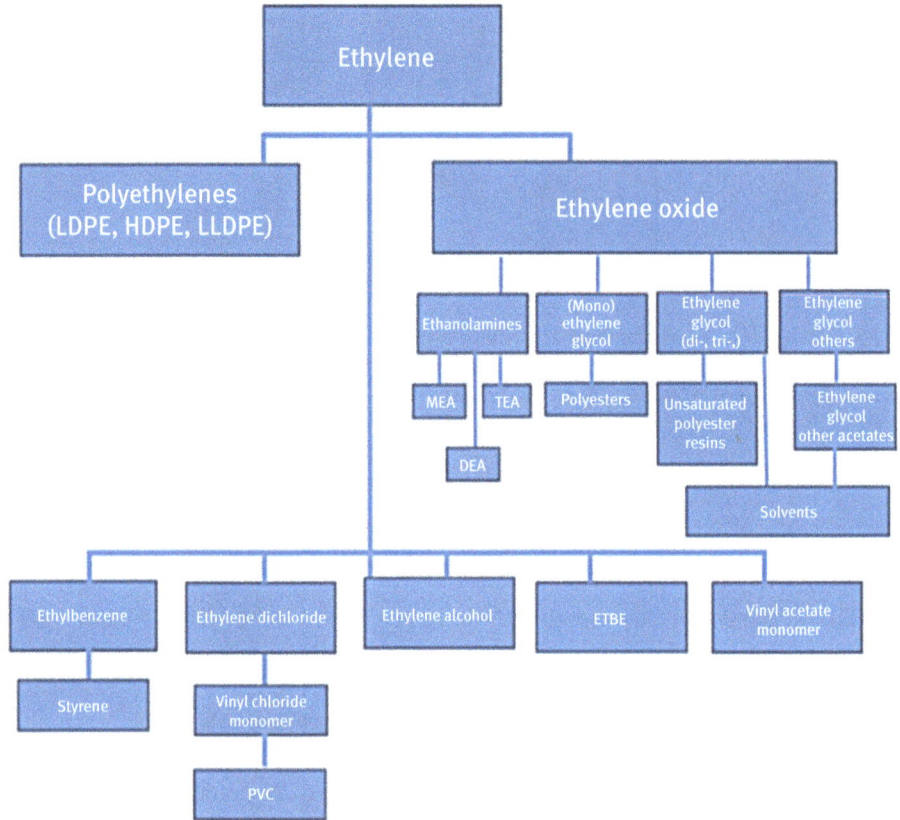

Figure 8.4: Some of the major petrochemical products from ethylene. (adapted from www.icis.com/ex plore/resources/icis-petrochemicals-flowchart/).

the *Clean Air Act* in the UK and USA) quickly spread across Europe and more recently around the world.[5]

Automobile emissions include the gases: N_2, CO_2, H_2O (which are not harmful in themselves); **CO**, **NO$_x$** (principally NO and NO_2); and **VOCs** (volatile organic compounds) produced mainly from unburned fuel in internal combustion chambers [7]. These toxic automobile exhaust gases may also undergo photochemical reactions in the atmosphere to produce even more harmful materials.

The **automotive catalytic converter** (ACC) has been the solution to this serious pollution problem. The ACC is placed in the exhaust-gas line ahead of the muffler as shown in Figure 8.5.

---

5 The Saudi Arabian Standards Organization (www.saso.gov.sa) decreed in the year 2000 that all car imports to the country must be equipped with Automobile Catalytic Converters. Domestically sold gasoline has been mixed with MTBE since then.

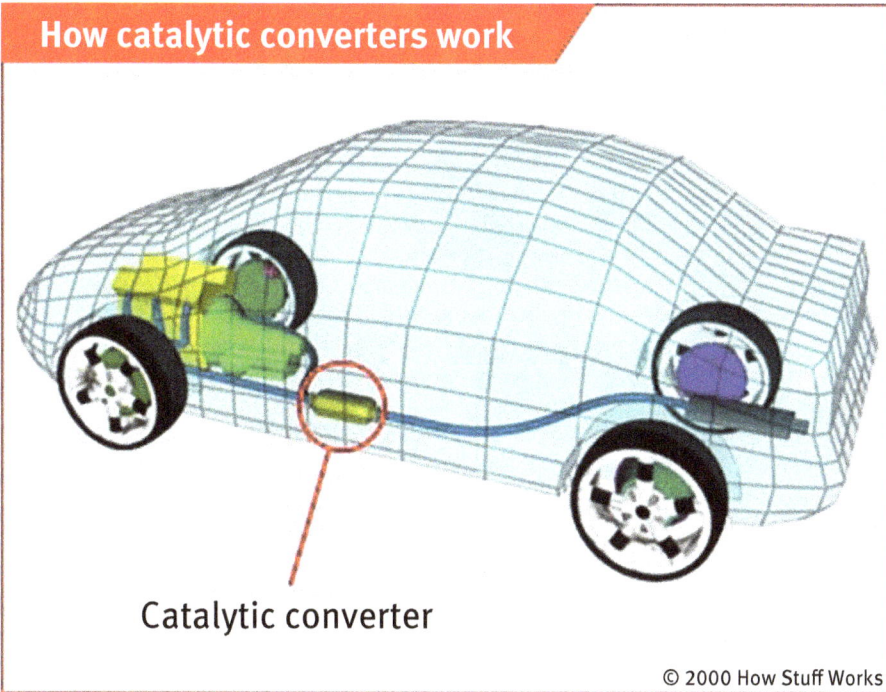

How catalytic converters work

Catalytic converter

© 2000 How Stuff Works

Figure 8.5: The automotive catalytic converter in the tail exhaust pipe.

Amazingly, the ACC works efficiently to reduce the amount of toxic pollutants to acceptable limits. For example: 5% VOCs are reduced to <0.2%; 1% CO is reduced to <0.05%; and 0.1% NO is reduced to <0.02%.

The residence time in the ACC is around 0.05 s (since the reactor volume is very small and the flow rate is extremely high). Considering those percentages, we see that conversions of around 95% are possible in the ACC.

In my opinion, the ACC is an example of a great chemical-engineering technology!

Figure 8.6 shows a photograph of an old ACC cut sideways to show the external and internal structures.

Figure 8.7 shows a schematic of the external and internal structure of a typical ACC. The internal framework of the converter is a ceramic *monolith honeycomb structure*, through which the air-gasoline combustion products flow.

The framework is designed to be strong and both impact-resistant and thermal-shock resistant. This is essential since, upon starting the engine, the converter can change temperatures by 500 °C in a very short period!

The active (precious) metals in the catalyst typically include Pt, Pd and Rh. These metals are deposited on the walls of the monolith coated with γ-Al_2O_3. The catalyst has been called a **three-way catalyst**.

Figure 8.6: An actual automobile catalytic converter.

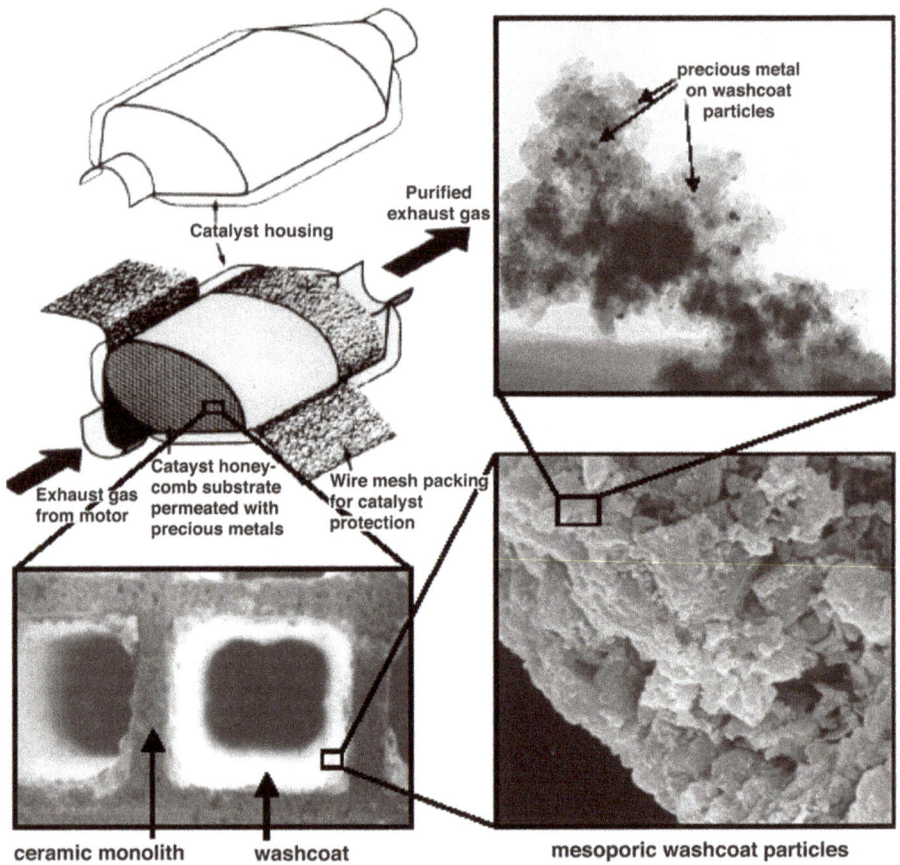

Figure 8.7: A schematic illustration of an automobile catalytic converter (ACC) with expanded views of the monolith framework, support, and active phase. (Mea culpa: I have misplaced the source of this schematic and the photographs).

Based on this name, the catalyst has three functions: The catalyst is supposed to convert the three main pollutants (CO, NO_x, and hydrocarbons) according to the following reactions:

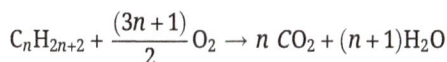

$$2\,CO + O_2 \rightarrow 2\,CO_2$$

$$2\,NO \rightarrow N_2 + O_2 \text{ and}$$

$$2\,NO_2 \rightarrow N_2 + 2\,O_2$$

$$C_nH_{2n+2} + \frac{(3n+1)}{2}O_2 \rightarrow n\,CO_2 + (n+1)H_2O$$

These reactions are mainly carried out by **Pt**, which is an excellent **oxidation** catalyst (recall the classification of catalysts by function in Chapter 1). However, the conversion of NO is more difficult, since it involves **reduction.** This is achieved by a number of reactions, as demonstrated below. **Rh** is an excellent metal for such reactions and acts to suppress side reactions that produce, for instance, N_2O.

$$CO + NO \rightarrow CO_2 + (1/2)N_2$$

$$H_2 + NO \rightarrow H_2O + (1/2)N_2$$

Stricter environmental laws are continually forcing car manufacturers to reduce automobile emissions. Indeed, the state of California, USA, is enacting laws that would require zero-emissions in the near future. The challenge of designing more efficient catalytic converters continues, despite the fact that this technology has existed for several decades. The chemical engineer in particular has played the most significant role in the development of ACCs.

Of course the slow but steady growth of *electric* cars, and to a lesser degree *H2-fueled* cars, is the future of the automobile industry. Most industrialized countries have already set 2030–2035 as the target years of the ban on producing fossil-fuel (gasoline and diesel) vehicles

References

[1] Silvy R. P. "Future Trends in the Refining Catalyst Market," Appl. Catal. A: Gener., 261, 247 (2004).
[2] https://www.grandviewsresearch.com/industry.analysis/catalyst-market
[3] https://www.saudiaramco.com/-/media/images/annual-review-2017/pdfs/en/2017-annualreview-full-en.pdf
[4] Morrison R. T., and Boyd R. N. "Organic Chemistry," Allyn and Bacon, Boston, 1973.
[5] Rase H. F. "Handbook of Commercial Catalysts: Heterogeneous Catalysts," CRC Press, Boca Raton, (2000).
[6] Schmidt L. D. "The Engineering of Chemical Reactions," Second Ed., Oxford University Press, New York, (2005).
[7] Hester R. E., and Harrison R. M. eds., "Volatile Organic Compounds in the Atmosphere," Royal Society of Chemistry, London, (1995).

9 Computational case studies

> There is great interest [in industry] on the use of mathematical models as a substitute for expensive experimentation.
> D. Ramkrishna and N.R. Amundson

The quotation under the title of this chapter appeared some 20 years ago in a review paper [1] on the role of mathematics in chemical engineering research, industrial innovation, and education.

Where are we now and how much progress has been accomplished?

The answer is **Industry 4.0**! This term is shorthand for the fourth industrial revolution.[1] Industry 4.0, in the context of the chemical industry, is perhaps easily understood by the cycle shown in Figure 9.1.

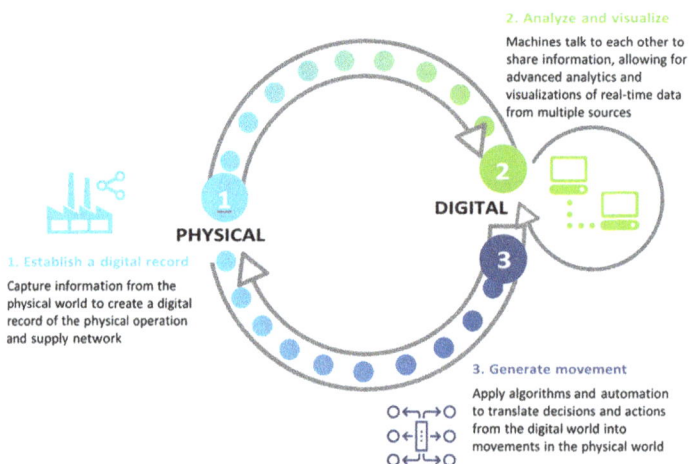

2. Analyze and visualize
Machines talk to each other to share information, allowing for advanced analytics and visualizations of real-time data from multiple sources

DIGITAL

PHYSICAL

1. Establish a digital record
Capture information from the physical world to create a digital record of the physical operation and supply network

3. Generate movement
Apply algorithms and automation to translate decisions and actions from the digital world into movements in the physical world

Figure 9.1: The physical-to-digital-to-physical leap of Industry 4.0 [2].

A key component of the cycle is "Analyze and visualize," under which we have *advanced analytics*. The components of advanced analytics include *data mining*; *modeling*; *simulation*; and *optimization*.

The objective of this chapter is to focus on two of those components and present few numerical case studies involving *modeling* and *simulation* in solid catalysis. This will be done keeping in mind that, nowadays, computational methods are very powerful and user-friendly, which is perhaps more important for students. The case studies are all relevant to the material presented in Chapters 1–8.

---

1 Briefly, the other industrial revolutions are (1) steam and mechanization, (2) science and mass production, and (3) digital technology and automation.

https://doi.org/10.1515/9783111032511-009

Let us briefly explain what the two terms, *modeling* and *simulation*, mean. This is especially important since "modeling and simulation capabilities are becoming an essential element of the **IoT** toolbox" [3]. IoT is shorthand for the **Internet of Things.** Students of engineering, science, or technology (dare I say lay persons?) are expected to know the meaning of IoT.[2]

Again, any student of engineering, science, or technology must have heard about the terms: modeling and simulation. It is fair, however, to say that few students know why modeling and simulation are essential elements in Industry 4.0, in IoT, and in scientific discovery. Figure 9.2 explains the role of modeling and simulation.

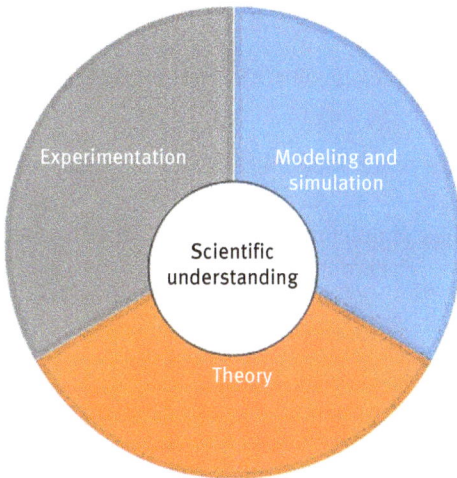

Figure 9.2: The role of modeling and simulation in scientific discovery.

Scientific understanding or discovery is based on formulation of "theory" to explain observed or measured phenomena; running "experiments" to test theory; and using the feedback of experimental results to refine theory.

The "model" is the mathematical expression of theory. "Simulation" is coding the model using a programing language: examples of languages include Mathcad®; MATLAB®; Mathematica®; FORTRAN, etc. These languages are what we nowadays call *software.*

9.1 Published models and simulators

Simulators are codes or programs enabling a computer to numerically solve mathematical models.

---

2 The reader is encouraged to visit the link at [5] for a short description of the history of the term IoT.

Perhaps the largest number of open-access simulators, related to engineering in general and catalysis in particular, are the **Wolfram Demonstration Projects** which are available at https://demonstrations.wolfram.com. In this section we will look at a few of those demonstration projects.

Ideally, the demonstrations are presented and interacted with in computer labs or in smart classrooms. Make sure you download the latest version of the **Wolfram Player,** available at https://www.wolfram.com/player/ before you start running the projects.

9.1.1 Adsorption isotherms of C_2H_4 in NaX zeolite

http://demonstrations.wolfram.com/AdsorptionIsothermsOfEthyleneInNaXZeoliteStructureFAU/

We have discussed adsorption isotherms: Langmuir, Freundlich, BET, etc. in Chapter 2. The purpose of this demonstration project is to use a form of the BET isotherm to simulate the adsorption of ethylene on NaX zeolites at different pressures and temperatures. The demonstration is straightforward. Some results are shown in Figure 9.3.

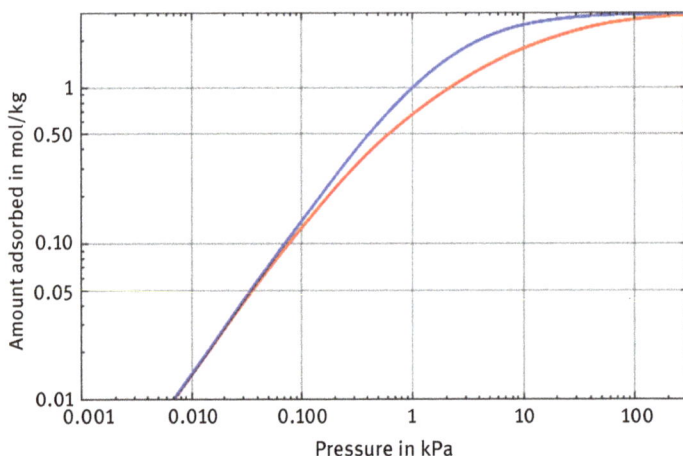

Figure 9.3: The adsorption of C_2H_4 on NaX zeolite using the BET isotherm (red) and Langmuir isotherm (blue).

It is instructive to go back and look at the isotherm given in this demonstration and to see how it is related to the general or standard forms of the BET isotherm.

Likewise, it is useful to set the constants C_j, $j = 1$–3, equal to zero to confirm whether indeed we obtain the Langmuir isotherm.

9.1.2 Surface kinetics with pore diffusion resistance

http://demonstrations.wolfram.com/SurfaceKineticsWithPoreDiffusionResistance/

This demonstration considers a special case of the problem we have solved in Section 5.2.1 of Chapter 5: a first-order reaction taking place in a porous catalyst particle with negligible external mass transfer resistance (i.e., the case where $Bi \gg 1$).

The demonstration allows you to look at the effect of the Thiele modulus on the concentration profile within the particle and the value of the effectiveness factor (called E here). An example of the concentration profile is shown in Figure 9.4.

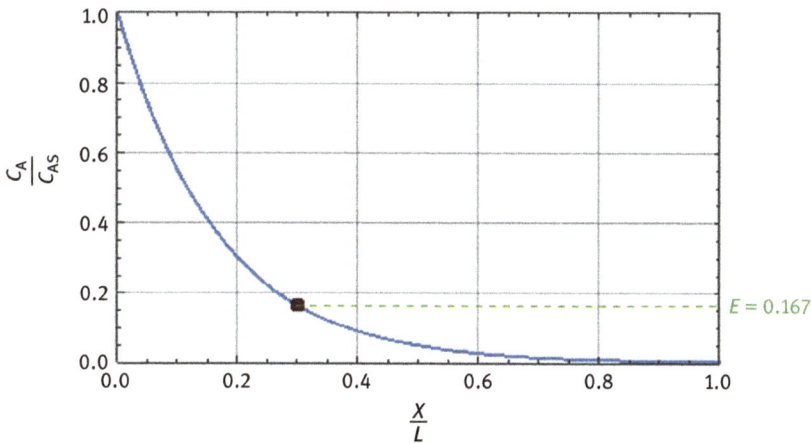

Figure 9.4: Concentration profile in a catalyst particle; exothermic first-order reaction in a slab-like catalyst particle.

It is instructive to compare the concentration profiles in this demonstration with those presented in Chapter 5. The difference you may see is a result of the *normalization* of the model equations. In Chapter 5, $x = 0$ and $x = 1$ represent the center and surface of the particle, respectively; whereas in the demonstration, $x = 0$ is the surface and $x = 1$ is the center.

9.1.3 Non-isothermal effectiveness factor

http://demonstrations.wolfram.com/NonisothermalEffectivenessFactorTheWeiszAnd HicksProblem/

In Chapter 5, we derived and looked at expressions for the effectiveness factor using different geometries: there are three expressions for η in Table 5.1 whose derivations are based on *isothermal* first-order catalytic reactions.

In this demonstration, we consider the *non-isothermal* case which has been al-luded to in Chapter 5 (see the footnote in page 69). The mole and energy balances are given in the demonstration for spherical geometry. In dimensionless form,[3] we have a description of the situation in terms of the following boundary-value problem:

$$\frac{d}{d\bar{r}}\left[\bar{r}^2\frac{da}{d\bar{r}}\right] - \bar{r}^2\phi^2\exp\left[\frac{\gamma\beta(1-a)}{1+\beta(1-a)}\right]a = 0$$

$$\frac{da}{d\bar{r}}(0) = 0$$

$$a(1) = 1$$

Note that $a \equiv C_A(r)/C_{A0}$ and $\bar{r} \equiv r/R$. Unlike the isothermal problem (with negligible ex-ternal transport effects), we now have *three* dimensionless parameters (γ, β, ϕ). These groups are defined in the demonstration: γ represents the magnitude of the activation energy; β the intensity (and type) of enthalpy of reaction; and ϕ the Thiele modulus.

In this demonstration, you can choose $\gamma = 1, 5, 10, 15$, or 30. The demonstration dis-plays the effectiveness factor (red curve), a measure of the total reaction rate inside the particle compared to its value at the particle surface, versus the Thiele modulus, for various values of the other parameters.

An example of the results is shown in Figure 9.5, where we have more than one solution for $\eta = f(\phi; \gamma, \beta)$. This phenomenon is called *multiple steady states*. The

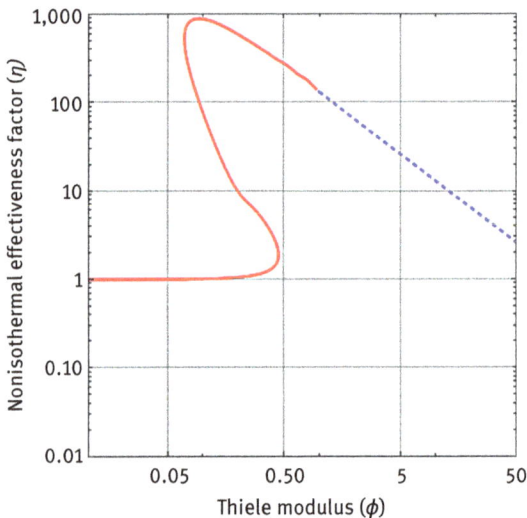

Figure 9.5: Effect of Thiele modulus on effectiveness factor for an exothermic first-order reaction in a spherical catalyst particle.

---

3 To maintain consistency, I have retained the same symbols used in Chapter 5.

demonstration also allows you to examine the effect of these parameters on the concentration and temperature inside the spherical catalyst particle, that is, $\bar{r} \in [0,1]$. Note that *negative* values of the parameter β, by definition, imply endothermic reactions.

9.1.4 Conversion of ethylbenzene to styrene in the presence of steam

http://demonstrations.wolfram.com/ConversionOfEthylbenzeneToStyreneInThePre senceOfSteam/

Today styrene is one of the most important monomers produced by the chemical industry. It is a basic building block in the plastics industry. The conventional method of producing styrene involves the catalytic alkylation of benzene with ethylene to produce ethylbenzene, followed by the *catalyzed* (by Fe-Cr oxides) dehydrogenation of ethylbenzene to styrene. The reaction is carried out in the presence of steam because the reaction is highly *endothermic*:

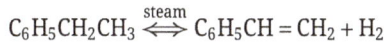

$$C_6H_5CH_2CH_3 \overset{\text{steam}}{\Longleftrightarrow} C_6H_5CH = CH_2 + H_2$$

Styrene undergoes polymerization by the common technologies used in the plastics industry to produce a wide variety of polymers and copolymers. *TotalEnergies, Shell, LyondellBasell, Formosa*, Saudi Arabian Petrochemical Co. (*SADAF*), and *Americas Styrenics*, are some of the leading companies active in the worldwide styrene market.

In this demonstration, the equilibrium constant $K_{eq} = f(T)$ for the dehydrogenation reaction is used to calculate the *equilibrium* conversion (the maximum possible) of ethylbenzene and the effect of total pressure and (steam/ethylbenzene) flow rate ratio. An example of the behavior of the equilibrium conversion is shown in Figure 9.6.

As an exercise, the reader is encouraged to derive an expression for the equilibrium conversion, where K_{eq}, pressure, and flow rate ratio appear. In fact, this expression can be used to check the consistency of the results with Le Chatelier's principle.

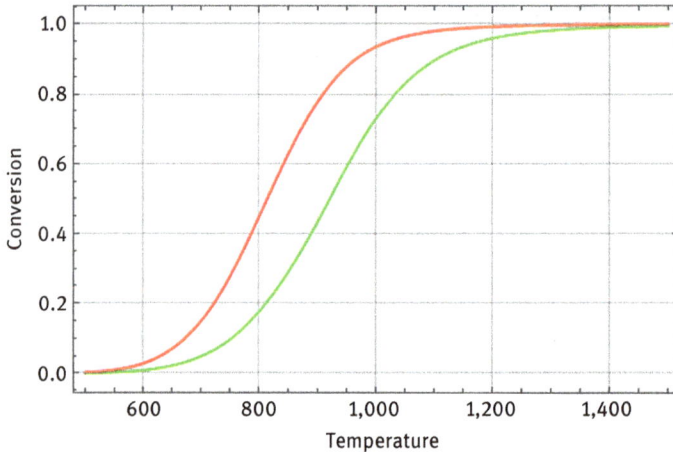

Figure 9.6: Effect of temperature on the equilibrium conversion of ethylbenzene to styrene in the presence (red curve) and absence (green curve) of steam.

9.1.5 Pressure drop in a packed-bed reactor

http://demonstrations.wolfram.com/PressureDropInAPackedBedReactorPBRUsingTheErgunEquation/

The design of isothermal PBRs was presented in Chapter 7. A key assumption in that analysis is that of negligible pressure drop across the length of the reactor.

In the present demonstration project, that assumption is relaxed and details including the Ergun equation in reactor calculations of two PBRs of lengths 14 and 21 m are presented for a first-order irreversible, gas-phase reaction.

The demonstration allows us to calculate the conversion and molar flow rate of the reactant, as well as the pressure along the reactor axis. A sample calculation is shown in Figure 9.7. Note that the effect of the particle diameter can also be studied.

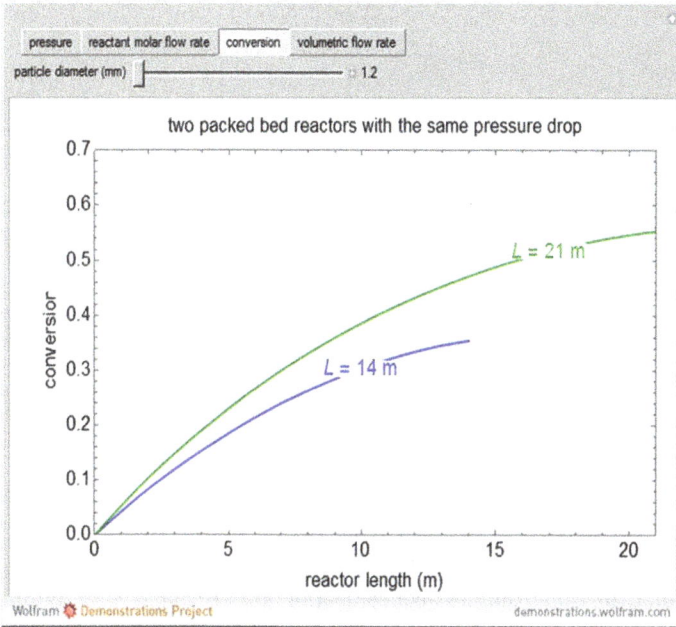

Figure 9.7: Conversion as a function of reactor length in two packed-bed reactors.

9.1.6 Slurry reactor with declining catalyst activity

http://demonstrations.wolfram.com/SlurryReactorWithDecliningCatalystActivity/

We have briefly looked at the physical characteristics of 3-phase catalytic reactors in Chapter 7: the trickle-bed and slurry reactors. We have also considered catalyst deactivation in Chapter 6. This demonstration is related to both chapters.

In this demonstration project, we focus on the mechanically agitated slurry reactor (Figure 7.7a) which could be considered a 3-phase CSTR. We are given a first-order reaction, A → B, taking place in this 3-phase CSTR. The catalyst suffers from deactivation, where the catalyst activity[4] follows the first-order relation:

$$\frac{da}{dt} = -k_d a$$

$$a\,(0) = 1$$

As shown in Figure 9.8, the demonstration calculates and plots the dimensionless concentration of the reactant versus time for user-set values of the residence time, deactivation constant, and reaction rate constant.

---

4 Recall that in Chapter 6 we use **a** for activity.

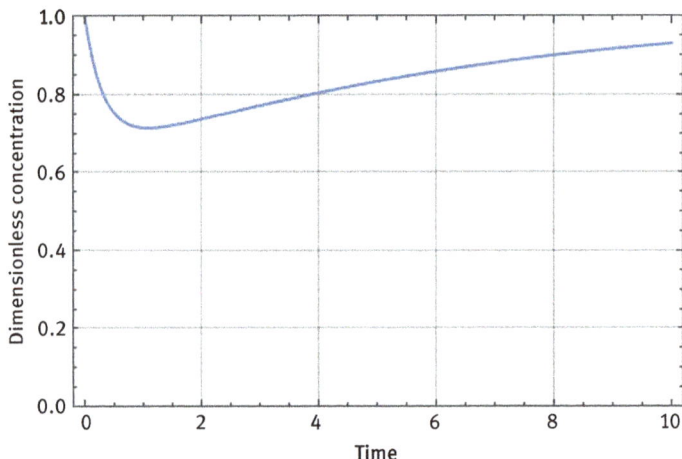

Figure 9.8: Change of reactant concentration with time in the slurry reactor.

The demonstration also allows us to observe the behavior of catalyst activity versus time. It is found that the dimensionless concentration goes through a minimum when there is deactivation, and it reaches a steady-state value when the activity is constant, as displayed in other "snapshots."

It is instructive for the reader to attempt to derive the model equation upon which the results in this demonstration are based. Therefore, the reader should start by stating the relevant assumptions, derive the mole balance for the reactant, and find an analytical solution, if possible, or a numerical solution leading to $[C_A/C_{A0}] = f(t, k, a, \tau)$, where τ is residence time. Once that is done, it would be interesting to see how the product concentration changes with time.

9.2 Applied catalysis design problem

Ethylene oxide is an important **specialty** chemical required for the manufacture of many important products, such as ethylene glycol (the typically green anti-freeze/anti-boil solution mixed with water in the car radiator); polyester (used in clothes, curtains, activewear, etc.); and (c) polyethylene terephthalate or PET (a thermoplastic polymer that is used to make liquid bottles and food containers, etc.).

Therefore, many petrochemical companies include ethylene oxide among their product portfolio; for example, *Dow Chemical, SABIC, BASF,* and *Sadara/Saudi ARAMCO.*

In this section we shall look at *model* equations and *simulation* of an isothermal PBR producing ethylene oxide.

The production of ethylene oxide is done by the silver-catalyzed (see Figure 9.9) vapor-phase oxidation of ethylene with air:

$$C_2H_4 + \frac{1}{2}O_2 \xrightarrow{\text{Ag/Al}_2O_{3(s)}} C_2H_4O \qquad (9.1)$$

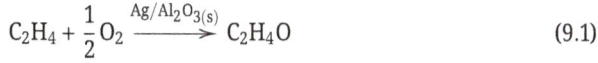

However, other reactions are possible where ethylene is totally oxidized and/or ethylene oxide is further oxidized:

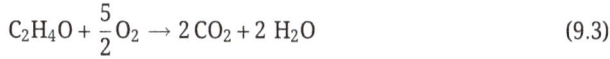

$$C_2H_4 + 3\,O_2 \rightarrow 2\,CO_2 + 2\,H_2O \qquad (9.2)$$

$$C_2H_4O + \frac{5}{2}O_2 \rightarrow 2\,CO_2 + 2\,H_2O \qquad (9.3)$$

In this process, as mentioned in Chapter 1, few ppm of *ethyl chloride* are added to the gaseous feed to inhibit the complete oxidation reaction (9.2) and increase the selectivity to ethylene oxide. Moreover, small traces of *byproducts*, namely acetaldehyde and formaldehyde, are also produced during the reaction.

Figure 9.9: Typical Ag/a-Al$_2$O$_3$ industrial catalyst for ethylene oxide (www.environmental-expert.com/companies/cri-catalyst-company-cricc-35454/).

Reactions (9.1)–(9.3) are exothermic and heat effects within and around the catalyst particles could be important. However, in this design problem, following Fogler [4], we will assume *isothermal operation*; *ignore the side reactions* (9.2) and (9.3); and we will further assume that *mass-transport limitations are negligible*.

Our problem then is given as follows: **Given the data in** Table 9.1, **calculate and graphically show how the conversion of ethylene, pressure, volumetric flow rate, and reaction rate change with the catalyst weight.**

Table 9.1: Data for design of PBR for production of ethylene oxide [4].

Feed	C_2H_4 (A) and O_2 (B) are fed in stoichiometric ratio. $F_{A0} = 136.21$ mol/s $p = 10$ atm
Reaction	$(-r_A') = k\,P_A^{1/3}P_B^{2/3}$ k (260 °C) = 0.00392 mol/atm. kg cat. s
Reactor	PBR: 10 banks of ½-inch diameter, schedule-40 tubes packed with catalyst; each bank has 100 tubes. Isothermal operation at 260 °C

Catalyst	Particle diameter = ¼ inch Density = 1,925 kg/m^3 Void fraction = 0.45
Physical properties of gas mixture	Identical to air

Obtained from FOGLER, H. SCOTT, ELEMENTS OF CHEMICAL REACTION ENGINEERING, 5th Ed., ©2016. Reprinted by permission of Pearson Education, Inc., New York, New York.

The model equations required to solve this design problem include: the mole balance for component A; reaction rate equation; stochiometric relationships; and the Ergun equation which relates pressure to the weight of the catalyst, as we have discussed earlier.

The model equations, in their final form are summarized in Table 9.2. Details of how those equations were derived are given in example 5–7 in Fogler [4].

Table 9.2: Summary of model equations for the PBR.

Mole balance	$\dfrac{dx}{dW} = \dfrac{(-r'_A)}{F_{A0}}$
Reaction rate law	$(-r'_A) = k'\dfrac{(1-x)}{(1+\varepsilon x)}p$
Stoichiometric relations	$C_A = \dfrac{C_{A0}(1-x)}{(1+\varepsilon x)}\left(\dfrac{p}{p_0}\right)$ $C_B = \dfrac{\left(\theta_B - \frac{x}{2}\right)}{(1+\varepsilon x)}\left(\dfrac{p}{p_0}\right)$
Ergun equation	$\dfrac{dp}{dW} = -\dfrac{a}{2\left(\frac{p}{p_0}\right)}(1+\varepsilon x)$
Constant parameters	$F_{A0} = 0.1362$ mol/s $k' = 0.0074$ mol/atm. kg cat. s $a = 0.0367$ 1/kg $\varepsilon = -0.15$ $\theta_B = 0.5$

A closer look at the model equations will show what we need to solve a set of *nonlinear* ordinary differential equations to simulate the reactor behavior. Therefore, we must use a numerical technique to generate solutions.

This has been done using the software MATLAB®. A summary of the code is given in Figure 9.10. The results are given in Figures 9.11a and b. The code can also be used to answer "what if" scenarios in the engineering design by suitable modifications. Herein lies the value of simulation.

As an illustration, you can use the same code with minor modifications to show the change of the reactants and product concentrations with catalyst weight.

FUNCTION FILE

```
function xprime = PackedBedReactorModel(W,x)


x1 = x(1);
x2 = x(2);


FA0 = 0.1362;
k = 0.0074;
Alpha = 0.0367;              % in kg^-1
expansion = -0.15;

ReactioRate = k * (1 - x1) / (1 + expansion * x1) * x2;

xprime(1,1) = ReactioRate / FA0;

xprime(2,1) = - Alpha * (1 + expansion * x1) / (2 * x2);
```

MAIN FILE

```
clear all
clc


k = 0.0074;
expansion = -0.15;


[W,y]=ode15s(@PackedBedReactorModel,[0 27],[0.0 1]);

ReactioRate = k * (1 - y(:,1)) ./ (1 + expansion * y(:,1)) .* y(:,2);

f = (1 + expansion * y(:,1)) ./ y(:,2);

figure(1)
plot(W,y(:,1),'r',W,y(:,2),'b',W,f,'g');
axis([0.0 27 0 3.0])
set(gca,'xtick',0:2.7:27,'ytick',0:0.3:3)
xlabel('W (kg)')
legend('x','p','f')


figure(2)
plot(W,ReactioRate,'b');
xlabel('W (kg)'), ylabel('(-r_A)')
axis([0.0 27 0 0.008])
set(gca,'xtick',0:2.7:27)
```

Figure 9.10: *MATLAB* code for solving the ODEs in Table 9.2 (courtesy of Dr. N.S. Abo-Ghander, KFUPM).

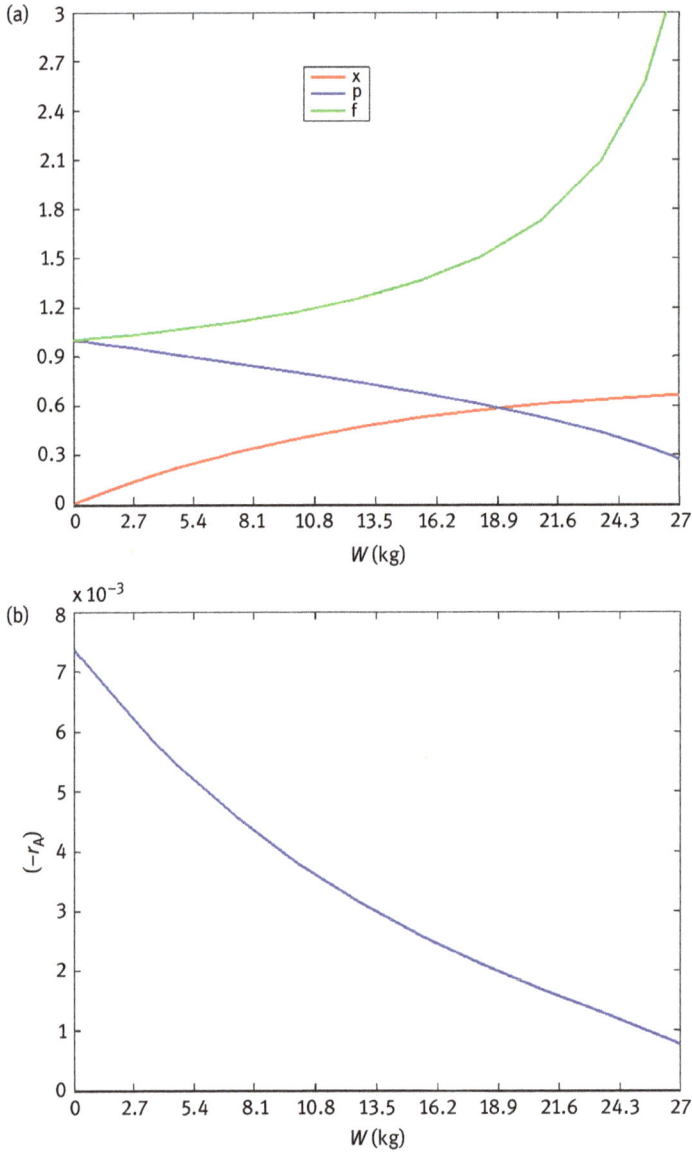

Figure 9.11: (a) Change of conversion (x) and dimensionless pressure (p) and volumetric flow rate (f) with catalyst weight. (b) Change of reaction rate ($-r'_A$) with catalyst weight (courtesy of Dr. N.S. Abo-Ghander, KFUPM).

9.3 Open-ended catalysis design problems

Let us go back and take another look at the assumptions made in solving the PBR design problem: isothermal operation, kinetics based on one reaction (reaction (1) only), and finally, negligible transport effects.

We will see that when *each* of these assumptions is relaxed, we can end up with an interesting "thought experiment."

It is useful to convert each of those experiments into a "computational thought experiment" by modifying the code given in Figure 9.10, repeating the computations, and studying the effect.

9.3.1 Elementary computational thought experiment

Let us start with considering the other reactions, (2) and (3), in the kinetics of the ethylene oxide reaction. You are required to:

(a) Use the following symbols, $A = C_2H_4$, $B = O_2$, $C = C_2H_4O$, $D = CO_2$, and $E = H_2O$, and write reactions (1)–(3) in this alternative form.

(b) The reaction rate constant k_1 is given in Table 9.1. Search the literature and try to find *approximate* values of the reaction rate constants k_2 and k_3 at the same operating temperature.

(c) Modify the rate expression $(-r'_A)$ in Table 9.1, to include reaction (2). Again, you need to search the literature to find the rate equation for this reaction.

(d) Modify the MATLAB code in Figure 9.10 to calculate the change of conversion, pressure, and volumetric flow rate with catalyst weight. Plot these changes and compare with the results shown in Figure 9.11a.

(e) You may still remember from your background in reactor design that, in the case of *multiple* reactions, yield and selectivity of the desired product C are more important than reactant conversion. Use the same code with suitable modifications to find the optimum catalyst weight to obtain the *maximum* yield of product C.

9.3.2 Intermediate computational thought experiment

In this section we will relax the third assumption by looking at internal and external transport effects but keeping the first assumption and the second assumption.

Therefore, we will focus on estimating the effectiveness factor. You are required to:

(a) Consult Table 9.1 and decide on what we need to do to evaluate the effectiveness factor η for reaction (1).

(b) Use your skills in numerical methods to solve the reaction–diffusion equations, presented in Chapter 5 to evaluate η.

(c)   What extra physical or chemical properties are required to find the value of η. Can we use the data given in Table 9.1 to estimate those parameters?

(d)   Use the value of η to modify the *MATLAB* code in Figure 9.10 and calculate the change of conversion, pressure, and volumetric flow rate with catalyst weight. Plot these changes and compare with the results shown in Figure 9.11a.

References

[1]   Ramkrishna D. and Amundson N. R. "Mathematics in Chemical Engineering: A 50 Year Introspective," AIChE J., 50 (1), 7 (2004).

[2]   Van Thienen S., Clinton A., Mahto M., and Sniderman B. "Industry 4.0 And the Chemicals Industry: Catalyzing Transformation through Operations Improvement and Business Growth," Deloitte University Press, (2016).

[3]   https://www.automationworld.com/article/industry-type/all/internet-things-and-importance-modeling-and-simulation; (August 2015).

[4]   Fogler H. S. "Elements of Chemical Reaction Engineering," Fifth Ed., Prentice Hall, Boston, MA, (2016).

[5]   https://internetofthingsagenda.techtarget.com/definition/Internet-of-Things-IoT

Problems

This is an hour for reflection.
James J. Carberry

Notes: Some of the problems have been adapted (with modifications) from various sources over several years. I have tried to cite all the original references, but in a few cases I lost track of the reference.

Problem 1: Visit this link and list the top five catalyst companies in the world. Arrange in descending order of the solid catalyst market share:
   Top 5 Catalyst Companies in the World | Expert Market Research

Problem 2: The oxidation of NH_3 was mentioned in Chapter 1 as an example of an industrial reaction where the catalyst is "unsupported" Pt gauze. Why is this reaction important?

Problem 3: Alhassani et al. (*IJE Trans. B: Applications,* **31** (8), 1172 (2018)) recently studied the kinetics of the following esterification reaction in which an Amberlyst-15 catalyst was used in a stirred batch reactor:

$$CH_3COOH + C_2H_5OH \rightarrow C_4H_8O_2 + H_2O$$

Using power-law kinetics, the orders with respect to ethyl acetate (A) and ethanol (B) were found to be 0.92 and 0.88, respectively.
   Use the following *initial-rate data,* collected in that study at 70 °C, to determine the value of the reaction rate constant.

C_{A0} (mol/L)	1	2	3	4	1	1	1
C_{B0} (mol/L)	1	1	1	1	2	3	4
$-r_{A0}$ (mol/L min) × 10^3	4	8.33	11.90	13.88	8.06	11.76	12.98

Problem 4: MnO_2 catalyzes the thermal decomposition of $KClO_3$ as follows:

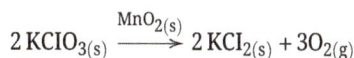

$$2\,KClO_{3(s)} \xrightarrow{\;MnO_{2(s)}\;} 2\,KCl_{2(s)} + 3O_{2(g)}$$

Both compounds are solid. Is this an example of homogeneous or heterogeneous catalysis? Justify your answer.

Problem 5: The following results were reported by Langmuir for the adsorption of nitrogen on mica at 20 °C:

https://doi.org/10.1515/9783111032511-010

Pressure (atm)		2.8	4.0	6.0	9.4	17.1	33.5
Amount of gas adsorbed(mm³ at 20 °C, 1 atm)		12.0	15.1	19.0	23.9	28.2	33.0

a) Plot the data to test whether the Langmuir isotherm is a good representation. If it applies, evaluate the adsorption constant.

b) A maximum of $\approx 10^{15}$ molecules usually cover 1 cm² of a surface. Estimate the effective surface area in Langmuir's experiment.

Problem 6: Derive a general expression for surface coverage θ_A where A is adsorbing on the surface in the presence of i different species ($i > 2$), all obeying the Langmuir isotherm.

Problem 7: Modify the Langmuir isotherm to allow for the *dissociation* of the molecule into *two* parts upon adsorption. *Hint*: dissociation means a molecule will occupy two sites on adsorption. Therefore, we can represent the situation in this case as an elementary dynamic equilibrium step as follows:

$$A_2 + 2\,X \Leftrightarrow 2\,(AX)$$

Problem 8: Given the following data for physical adsorption of nitrogen on alumina at −195 °C, calculate the total surface area using the BET method:

P/P_0	0	0.01	0.08	0.15	0.3	0.45
V (cm³/g) at STP	0	75	100	120	40	160

Problem 9: Measurements for the adsorption of NO_2 on charcoal were made by McBain and Britton in 1930, in terms of P vs. W, where W is weight of NO_2 adsorbed per gram of charcoal:

P (atm)	0.6	4.0	9.7	19.8	38.0
W (g/g)	0.1525	0.1891	0.1971	0.2064	0.2079

Does the Freundlich isotherm represent these data accurately? Support your answer by a plot using the linear form of this isotherm.

Problem 10: The following relationship between surface area and mean pore radius for cylindrical pores was given in Chapter 3. Derive this relationship.

$$\bar{r}_p = \frac{2\,V_g}{S_g}$$

Problem 11 (data extracted from microtrac-BEL.com):
The graph below shows pore volume distribution using nitrogen. What is the value of the mean pore radius and what type is it?

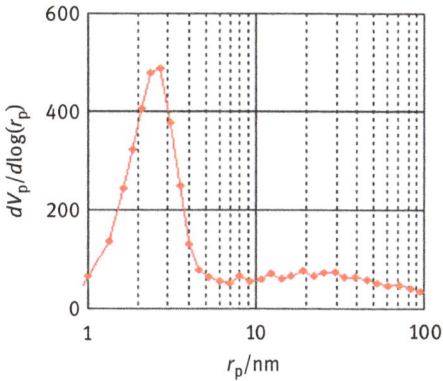

Problem 12: Consider the catalytic reaction between species A_2 and B. Derive an L-H rate expression if one of the molecules (e.g., species A_2) dissociates into equal fragments upon adsorption. The reaction mechanism can be expressed as:

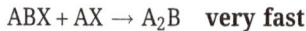

$$A_2 + 2X \Leftrightarrow 2AX$$

$$B + X \Leftrightarrow BX$$

$$AX + BX \rightarrow ABX \quad \textbf{slow}$$

$$ABX + AX \rightarrow A_2B \quad \textbf{very fast}$$

Problem 13: On a palladium catalyst surface, acetylene C_2H_2 undergoes five competing interactions:
(a) Reversible desorption
(b) Reversible C–H bond breakage
(c) Decomposition to carbon and hydrogen
(d) Hydrogenation to ethylene (no H_2 added!)
(e) Trimerization to benzene

Suggest mechanisms based on these interactions.

Problem 14: Derive the concentration–distance relationship for a first-order catalytic reaction in a porous *cylindrical* pellet, $a = f(z; \phi, Bi)$. Use the relationship to investigate the effect of ϕ and Bi on the concentration profile. Plot concentration profiles using the following parameter values:

Figure 1: Bi = 1, and ϕ = 0.1, 1.0, 5.0

Figure 2: ϕ = 1, and Bi = 0.01, 10, and 100

Briefly discuss the sensitivity of the profiles to these two groups.

Problem 15: Consider the effectiveness factor for *spherical* catalyst particles in the case of first-order kinetics. Investigate the asymptotic behavior of this expression at large and small values of the Biot number and Thiele modulus.

Problem 16: Consider the cracking of cumene C_9H_{12} by means of silica-alumina catalyst particles at 1 atm and 510 °C. Use the following data to calculate the effective diffusivity, and determine the dominant mode of diffusion:

    Particle diameter = 0.43 cm

    Specific surface area = 342 m^2/g

    Particle density = 1.14 g/cm^3

    Porosity = 0.51

    Tortuosity factor = 3

    Molecular diffusivity = 0.15 cm^2/s.

Problem 17 (modified from Bond, *Heterogenous Catalysis: Principles and Applications*, 2nd Ed., Oxford University Press, Oxford, 1987):

The rate of oxidation of butadiene to maleic anhydride and carbon oxides over a vanadium-phosphorus-oxides (VPO) catalyst varies with temperature according to the following data:

Temperature (°C)	310	325	342	358	375	397	410	426
Rate × 10^3 (mol/g cat. h)	0.24	0.50	1.19	2.26	3.50	4.87	5.35	5.95

a) Examine the data to see if the reaction is mass-transport controlled? *Hint:* there is little information about the catalyst, adsorption, kinetics, etc. in this problem. Therefore, to "examine" the data, plot ln (rate) versus $1/T$, and check the behavior.

b) If your answer is yes, calculate the activation energies in the high- and low-temperature regions and comment on the difference in case it is significant.

Problem 18: A first-order irreversible catalytic reaction $A \rightarrow P$ is carried out at isothermal conditions, *in the absence of external transport effects*, in a Carberry reactor on two different size catalyst pellets. The pellets are spherical. The catalyst activity and pore structure of the pellets are identical. Therefore, the kinetic rate constant and effective diffusivity are identical in pellets of both sizes. The temperature, pressure and bulk reactant concentration are identical in both runs. The following data are obtained:

	Pellet diameter	Observed rate × 10^5
	(cm)	(mol/cm^{3}cat. s)
Run 1	1.0	3.0
Run 2	0.1	15.0

a) Estimate the Thiele modulus ϕ and effectiveness factor η for each particle. Are internal diffusion effects important in each pellet size?

b) What pellet diameter is needed to completely eliminate internal diffusion resistance at the temperature of these experiments (i.e., for $\eta \to 1$)?

Problem 19 (modified from Carberry, *Chemical & Catalytic Reaction Engineering*, McGraw-Hill, NY, 1976):

The reaction of CO and Cl$_2$ has been studied over an activated-carbon catalyst:

$$CO + Cl_2 \to COCl_2$$

Reaction rate control appears to be that of surface reaction between the adsorbed reactants.

(a) Using L-H kinetics, derive a rate expression based on the assumption that only Cl$_2$ and COCl$_2$ are strongly adsorbed.

(b) Use the data given below to test the rate expression derived in part (a).

Rate × 10^3	P_{CO}	P_{Cl_2}	P_{COCl_2}
4.41	0.406	0.352	0.226
4.40	0.396	0.363	0.231
2.41	0.310	0.320	0.356
2.45	0.287	0.333	0.376
1.57	0.253	0.218	0.522
3.90	0.610	0.113	0.231
2.00	0.179	0.608	0.206

Problem 20: The irreversible catalytic oxidation of carbon monoxide to carbon dioxide was studied experimentally in detail by McCarthy et al. (*J. Catal.*, **39**, 29 (1975))

(a) Using A≡CO, B≡O$_2$, and C≡CO$_2$, develop a general L-H mechanism for this reaction assuming all species are adsorbed without dissociation.

(b) McCarthy used excess oxygen and Pt/α-Al$_2$O$_3$ catalyst. The figure below shows the data collected in a Carberry reactor at 246 °C for the reaction rate change with the concentration of CO.

Derive rate equations for the above reaction assuming (1) O_2 is strongly adsorbed, and CO_2 is not adsorbed. (2) O_2 is weakly adsorbed, and CO_2 is not adsorbed.

Which of the two rate equations do you think gives a better explanation of the experimental data?

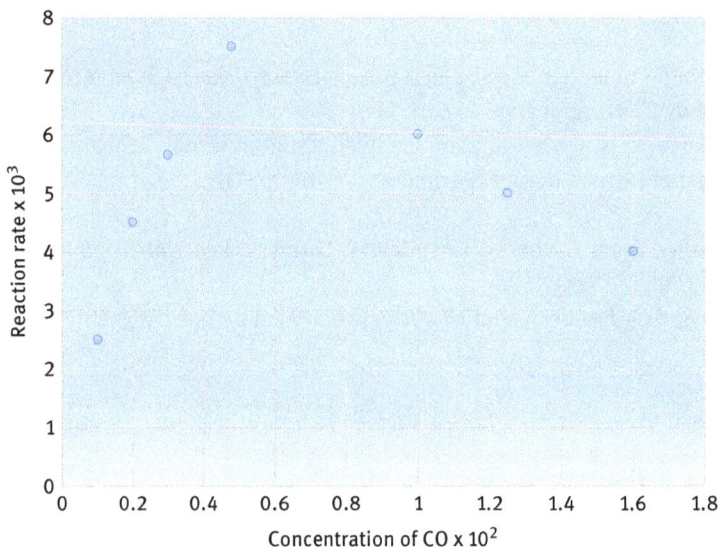

Concentration of CO x 10^2

Problem 21: Thodos and Stutzman (*Ind. Eng. Chem.*, **50**, 413 (1958)) studied the formation of ethyl chloride catalyzed using ZrO_2/silica gel. Methane was used in the experiments as an inert. The reaction is:

$$C_2H_4 + HCl \Leftrightarrow C_2H_5Cl$$

(a) If the kinetics are controlled by the surface reaction between adsorbed ethylene and adsorbed hydrochloric acid, develop an expression for the reaction rate.
(b) The authors reported the following experimental partial pressure-reaction rate data at 350 °F, where the feed contained different ratios of the reactants as well as methane. Use the data to evaluate the constants in the rate equation.

Run	P_{CH_4} (atm)	$P_{C_2H_4}$	P_{HCl}	$P_{C_2H_5Cl}$	Rate × 10^4 (lb mol C_2H_4/ h. lb cat.)
1	7.005	0.300	0.370	0.149	2.62
2	7.090	0.416	0.215	0.102	2.60
3	7.001	0.343	0.289	0.181	2.52
4	9.889	0.511	0.489	0.334	2.16
5	10.169	0.420	0.460	0.175	2.63

Problem 22 (adopted from Kandiyoti, R., *"Fundamentals of Reaction Engineering – Worked Examples,"* Ventus Publishing ApS, 2009):

The *differential reactor* is one of the laboratory reactors that we discussed in Chapter 5. It is used to simulate a very thin "slice" of the packed-bed reactor, and mainly used to measure reaction rates and parametric sensitivity, under the assumption of small conversions.

The first-order irreversible reaction A → B + C will be carried out over 2 g of spherical catalyst in a differential test reactor. The intrinsic reaction rate expression for the reaction in (kmol/kg cat. s) is:

$$-r_A = 7 \times 10^{15} \exp\left(-\frac{18,000}{T}\right) C_A$$

where T is in (K) and C_A is in (kmol/m^3).

Assuming isothermal operation at 500 K and given the data below,
(a) Estimate whether internal diffusion resistance is important.
(b) Estimate whether external mass transfer resistance is important.
(c) Calculate the conversion under these conditions.

Data

Property	Value
Catalyst pellet diameter	0.001 m
Catalyst pellet density	1,500 kg/m^3
Effective diffusivity	1 × 10^{-6} m^2/s
Molecular diffusivity	1 × 10^{-4} m^2/s
Mass flow rate	0.1 kg/m^2. s
Viscosity of gas mixture	1 × 10^{-6} kg/m.s
Density of gas mixture	1 kg/m^3
Void fraction of the bed	0.42
Total volumetric flow rate	0.005 m^3/s

Problem 23: Arab light gas oil (LGO) is cracked in a tubular reactor packed with silica-alumina spherical catalyst particles (average particle diameter = 1.62 mm). The liquid is vaporized, heated, and enters the reactor at 630 °C and 1 atm. With good temperature control, the reactor contents remain close to this temperature. For moderate conversions, the cracking reaction follows first-order kinetics with an intrinsic reaction rate constant of 52.53 s^{-1}. Given that the effective diffusivity is 8 × 10^{-8} m^2/s and the gas-solid mass transfer coefficient is 1 m/s, calculate the Biot number, Thiele modulus, and effectiveness factor. What can you conclude from these values about the mass transfer limitations?

Problem 24: The reaction of carbon monoxide and hydrogen is carried out on a nickel catalyst at 400 °C and 1 atm:

$$CO + 3H_2 \Leftrightarrow CH_4 + H_2O$$

At low conversions the rate of this catalytic reaction is given by the following expression:

$$-r_{CO} = \frac{1.1P_{CO}P_{H_2}^{\frac{1}{2}}}{1 + 1.5P_{H_2}} \quad \text{lbmol/lb cat. hr}$$

For an equimolar mixture of CO and H_2, fed to a packed-bed reactor at the rate of 30 lbmol/h, calculate the weight of catalyst required to obtain 20% conversion of CO. Mass transfer effects can be ignored.

Problem 25: For environmental reasons, it is proposed to reduce the concentration of nitric oxide in an effluent stream from a plant by passing it through a packed-bed reactor containing spherical particles of a carbonaceous porous catalyst. The feed stream consists of a mixture of 2% NO and 98% air flowing at the rate of 10^{-5} m³/s at 1,173 K and 101.3 kPa. The reaction rate law is:

$$-r_{NO} = kC_{NO}, \text{ where } k = 2.34 \times 10^{-7} \text{m}^3/\text{g cat. s}$$

Calculate the length of a 2 in-ID reactor tube which is necessary to reduce the concentration of NO to a level of 0.004%, i.e., to a value just below the maximum acceptable environmental limit.

Additional data:
Effective diffusivity = 2.66×10^{-8} m²/s
Bed density = 1.4×10^6 g catalyst/m³
Particle radius = 3×10^{-3} m
Gas-solid mass transfer coefficient = 6×10^{-5} m/s

Problem 26 (modified from *people.engr.ncsu.edu/jmhaugh/kinetics_probs.html*):
The first-order catalytic reaction A → B was studied in a PBR under isothermal conditions. Given the following data on the reactor and catalyst:
(a) Calculate the apparent rate constant, where $k_{app} = \eta k$.
(b) Determine whether the internal diffusion resistance is minimal, intermediate, or strong.

Reactor	Catalyst
Volume = 50×10^{-3} m^3	Weight = 80 kg
Fluid flow rate = 10^{-3} m^3/s	Density = 2.0×10^3 kg/m^3
Feed concentration of reactant A = 0.1×10^3 mol/m^3	Particle geometry: cylindrical, $d = 0.1$ cm, $H = 0.2$ cm
Exit concentration of reactant A = 0.055×10^3 mol/m^3	Effective diffusivity of reactant A = 5×10^{-6} cm^2/s

Problem 27: The catalytic reaction A → R was conducted in an isothermal batch reactor. The fluid volume = 10 L; catalyst weight = 5 kg, and catalyst density = 2.0 kg/L. The catalyst particles are rectangular slabs with thickness of 0.4 cm ($L = 0.2$ cm). In a separate kinetic study, the reaction has been found to follow *zero-order* kinetics under reaction conditions.

(a) Derive an expression for the effectiveness factor assuming no external mass transfer resistance.

(b) The plot below shows the concentration of the reactant as a function of time. Use the data to estimate the value of the rate constant (with catalyst weight as the basis).

(c) Estimate the value of the effective diffusivity of reactant A.

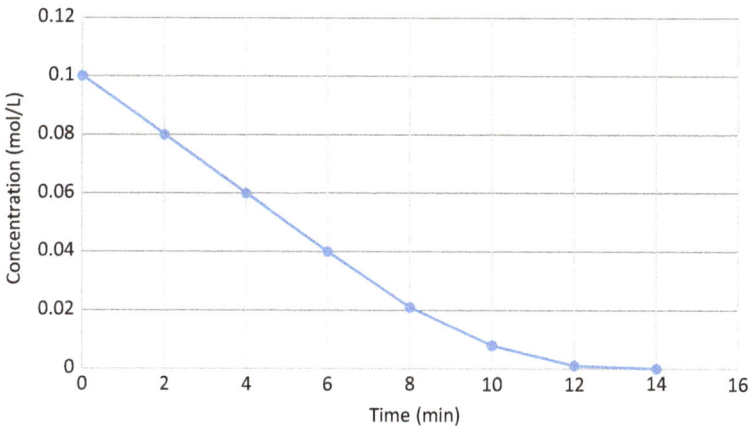

Problem 28: The esterification of maleic acid with ethanol in the presence of a cation exchange resin catalyst was studied by Sirsam and Usmani in a mechanically agitated reactor (*Int. J. Chem. Eng.*, **4** (4), 217 (2013)):

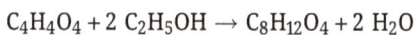

$$C_4H_4O_4 + 2\ C_2H_5OH \rightarrow C_8H_{12}O_4 + 2\ H_2O$$

The effect of external mass transfer on the reaction rate was studied by changing the agitation speed in the reactor at constant temperature. The data collected is shown below in terms of the conversion of maleic acid with time at different agitation speeds. What can you say about external mass transfer effects: minimal, intermediate, or strong?

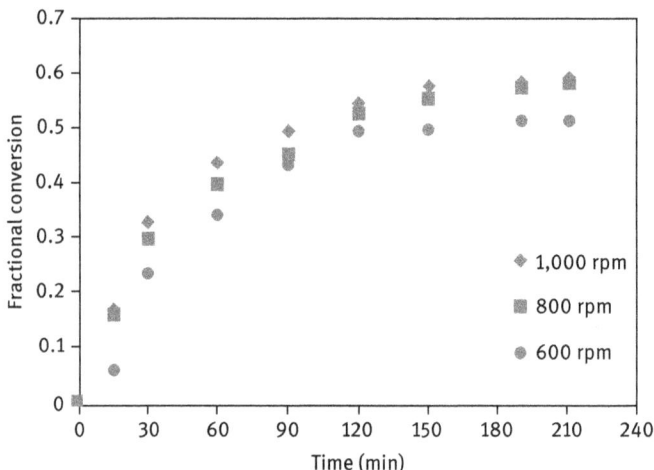

Problem 29: The hydrogenation reaction $A_{(g)} + 2\,H_{2(g)} \rightarrow R_{(g)}$ takes place on Ni/SiO$_2$ catalyst. The reaction is zero order in A and first order in H$_2$:

$$-r_A = k\,P_{H_2}, \text{ where } k = 1.74\,\text{mol/kg cat. bar. min}$$

The reaction takes place in a PBR, where the feed is 67% A and 33% H$_2$, at a total flow rate of 40 mol/min. The feed conditions are 70 °C and 5.0 bar. Moreover, the reactor contains 4.8 kg of catalyst.

(a) Assume the reactor operation is isothermal and isobaric, plot the molar flow rate of all three components as a function of catalyst weight.
(b) Given the following extra data to allow us to assess the pressure change along the reactor, repeat your calculations in part (a) and compare.

Additional data:
– The reacting fluid properties are identical to air.
– The density of the spherical ($d = ¼$ inch) catalyst particles is 1.925 kg/m^3; bed void fraction = 0.45; and gas density = 16 kg/m^3.

Problem 30: The Carberry reactor was used to obtain reaction rates for the solid catalyzed *i*-butane and styrene oxidations (Suppiah et al., *Progress in Catalysis*, **73**, 187–193 (1992)). The table below shows the catalyst activity (i.e., reaction rate) at different catalyst-basket rotation speeds using Pt/silicalite particles for the oxidation of *i*-C$_4$H$_8$ at 275 °C.

Rotation speed (rpm)	Catalyst activity (mol/g. min)
200	1.20
800	2.55
1,200	2.85
1,800	3.20
2,400	3.22
3,000	3.22

Plot activity vs. rotation speed. What can you conclude about mass transfer limitations at low and high speeds?

Problem 31: Carbon disulfide is a volatile liquid that nowadays is used primarily in the production of the packaging material "cellophane." In the MENA region, CS_2 is utilized in UAE and Saudi Arabia for that purpose. CS_2 is produced according to the following reaction catalyzed by silica gel or alumina:

$$CH_4 + 2\ S_2 \Leftrightarrow CS_2 + 2\ H_2S$$

Consider the production of CS_2 in an isothermal PBR at 600 °C operated at 1 bar. At 600 °C the reaction has been found to be independent of transport limitations. The reaction rate in (kmol/kg cat. s) is given by the following expression:

$$-r_{CH_4} = \frac{8 \times 10^{-2} \exp\left(-\frac{115,000}{RT}\right) P_{CH_4} P_{S_2}}{1 + 0.6\ P_{S_2} + 2.0\ P_{CS_2} + 1.7\ P_{H_2S}}$$

where the partial pressures are in (bar), the temperature in (K), and R in (kJ/kmol K).
(a) Suggest a mechanism that is consistent with the rate expression.
(b) Calculate the mass of catalyst required to produce 10 ton/day of carbon disulfide at a conversion of 75%, given the total inlet flow rate 3,373.5 mol/s in the stoichiometric ratio.

Problem 32: Toluene is mainly used as a precursor to benzene via hydro-alkylation. The second application involves its disproportionation to a mixture of benzene and xylene.

However, let us look at an interesting proposal to produce toluene by passing an equimolar mixture of benzene and xylene over a catalyst at 450 °C and 465 psia in an isothermal PBR:

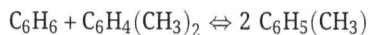

$$C_6H_6 + C_6H_4(CH_3)_2 \Leftrightarrow 2\ C_6H_5(CH_3)$$

The rate equation is given by the following expression where A is benzene, B is xylene, and C is toluene:

$$-r_A = \frac{k_1(p_A p_B - K p_C)}{(1 + p_A K_A + p_B K_{BA} + p_C K_C)^2}$$

where $k_1 = 0.86 \times 10^{-3}$, the reaction equilibrium constant reactant conversion $K = 3.5$, and the adsorption constants are $K_A = K_B = K_C = 0.362$

Given that the catalyst bulk density $= 45$ lb/ft^3, calculate the weight of catalyst required to convert 30% of 200 lb mol of the mixed feed. In the absence of relevant data, you can assume mass transfer effects are negligible.

Problem 33: (Inspired by the work of Grenoble, D.C. (*J. Catal.*, **56**, 40 (1979)) Hydrodealkylation is a cleavage reaction, where a H_2 molecule reacts with an organic compound, resulting in two smaller compounds. Consider the catalytic hydrodealkylation of toluene:

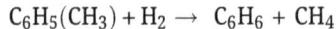

$$C_6H_5(CH_3) + H_2 \rightarrow C_6H_6 + CH_4$$

The rate equation at the reaction temperature is:

$$-r_A = 0.32 \, C_A C_B^{0.5} \text{ kmol/(m}^3 \text{ cat. s)}$$

where A is toluene and B is hydrogen. The effective diffusivity of toluene is 8.42×10^{-8} m^2/s. Is it possible to estimate the effectiveness factor when the external mass transfer effects are negligible? Suggest an alternative way to deal with this problem.

Problem 34: According to https://www.catalyticconverterrecycling.org/catalytic-converter-recycling-in-the-world/ automotive catalytic converters (ACC) are recycled and sold at very good prices! Look up and report the price *per gram*, in descending order, of the major metals in the ACC: Pt, Pd, and Rh.

Problem 35 (contributed by Dr. N.S. Abo-Ghander):
In an isothermal PBR, the gas phase reaction A $\rightarrow$ B occurs on slab-like catalyst pellets, where the feed flow rate of A is 0.1 mol/s and concentration is 0.5 mol/m^3:

$$-r_A = \frac{k \, C_A}{(1 + K_A C_A)} \text{ mol/(kg cat. s)}$$

Calculate the amount of catalyst required to achieve 80% conversion for three different cases: low (0.5), moderate (1.5), and high (5.0) Thiele moduli. The reaction rate constant is 0.1 s^{-1} and adsorption coefficient $K_A = 0.001$ m^3/mol.

You are also required to solve and compare your answers using the homogeneous model ($\eta = 1$) and first-level heterogeneous model (i.e., η must be estimated numerically using, e.g., the MATLAB function "bvp4c").

Index

abrasion 45–46, 88
abrasion resistance 46
ACC (automotive catalytic converter) 116, 118
acid catalyst 115
acidic solids 6
activated alumina 8, 36
activated carbon 8
activated carbons 26, 35
Activated clays 35
activation energy change 61
active agent 7, 10, 83, 87
active centers 7
active site 50
active sites 9, 15–16, 32, 53, 55, 64, 82, 84–86
activity 6, 9, 41, 81, 88, 91, 127–128
adsorbate 16, 26, 28–29, 33
adsorbent 20
adsorption 5
– heat 17, 20
– multilayer 27–28
– rate of 19, 21–23
– strength of 20, 22
adsorption 12, 15, 20, 26, 33, 41, 122
adsorption and TPD of bases 41
adsorption coefficient 24
adsorption constants 22, 53, 60, 136
adsorption equilibrium 22
adsorption site 22
aggregative fluidization 100
agitation speeds 144
air and water pollution 115
air pollution, automobile 1
alkylation 6, 111
ammonia oxidation 82
ammonia synthesis 81
applications, major 110
Arrhenius diagram 18
Arrhenius law 60
Arrhenius plot 61
ASTM 30
ASTM D-4,179 method 45
asymptotic behavior 138
attrition 83, 87, 88
attrition resistance 46
automobile emissions 115, 119

batch reactor 76
– stirred 135
bed cross-section 102
bed density 101, 104–105
bed height 45
bed porosity 104
bed-support media 92
BET isotherm 28, 30, 122
– standard 29
BET isotherm in linear form 30
BET method 33, 136
BFB reactor 100
bimolecular surface reactions 53
Biot number 67, 69–70, 138
Boudouard reaction 85
Bubble column slurry reactor 108
bubbling fluidized bed (BFB) 100
bulk bed density 101
bulk composition 41
bulk diffusion
– ordinary 73, 72
bulk diffusivity 74
bulk fluid 11, 63, 102
bulk phase 57
bulk structure morphology XRD 41

Carberry reactor 76–78, 138–139, 144
carbon dioxide 52, 56, 139
carbon disulfide 145
case studies, numerical 120
catalyst
– enzymatic 1
– homogeneous 1
– three-way 117
catalyst activity 9, 32, 81, 83, 88, 91
catalyst bed 88, 92, 95, 97, 102
catalyst deactivation 81–83, 85, 87–88, 103, 112, 127
catalyst pellet diameter 141
catalyst properties 32
catalyst recovery services 91
catalyst selectivity 6
catalyst shapes 70
catalyst synthesis 9
catalyst typical life 82

https://doi.org/10.1515/9783111032511-011

www.ingramcontent.com/pod-product-compliance
Lightning Source LLC
Chambersburg PA
CBHW081535220326
41598CB00036B/6444